INSIDE THE TECHNICAL
CONSULTING BUSINESS

INSIDE THE TECHNICAL CONSULTING BUSINESS

Launching and Building Your Independent Practice

THIRD EDITION

Harvey Kaye
Consultant in Mechanical Engineering
Rochester, New York

JOHN WILEY & SONS, INC

New York • Chichester • Weinheim • Brisbane • Singapore • Toronto

Copyright © 1998 by John Wiley & Sons, Inc.

This publication is designed to provide accurate and
authoritative information in regard to the subject
matter covered. It is sold with the understanding that
the publisher is not engaged in rendering legal, accounting,
or other professional services. If legal advice or other
expert assistance is required, the services of a competent
professional person should be sought.

Library of Congress Cataloging in Publication Data

Kaye, Harvey
    Inside the technical consulting business : launching and building
your independent practice / Harvey Kaye. — 3rd ed.
      p.   cm.
    Includes bibliographical references and index.
    ISBN 0-471-18341-5 (cloth : alk. paper)
    1. Consulting engineers.   2. Engineering firms — Management.
I. Title.
TA216.K39   1997
620'.0068'1 — dc21   97-22690

Printed in the United States of America
10 9 8 7 6 5 4

ENGINEERS, SCIENTISTS, AND TECHNICAL SPECIALISTS

This book shows how you can become a consultant in your technical field and enjoy professional independence and high income. If you are an engineer, a scientist, a technical writer, a computer programmer, a quality control expert, a mechanical designer, a patent expert—in fact, any kind of technical specialist—looking for career enhancement options, this book is for you! By following the tips and guidance in this book, you can start and manage your own consulting practice.

This book is not a frilly or eulogistic praise of consulting. It shows, in no-nonsense terms, what it takes to be a successful consultant. It gives you marketing, client relations, and self-management know-how that has taken the author twenty-five years to accumulate, and that is not in print anywhere else. The secrets of successful marketing are told clearly and concisely, so that you can concentrate on your best bets and minimize activities that have no payoff.

WHAT IS CONSULTING REALLY LIKE?

I give you a behind-the-scenes look that other consultants are too tight-lipped to share. You gain valuable insight into what it is like to be a consultant without leaving the comfort of your own chair!

Many successful consultants portray consulting as mysterious and nearly impossible, like climbing Mount Everest. This makes them look like geniuses, but it doesn't help you figure out how to go about it yourself. I show you, in simple language, how you can *build* on your present knowledge and abilities and *plan* your business growth. I can't do the work for you, but I can *illuminate* your own path to professional satisfaction as a consultant.

LEARN TO BREAK AWAY FROM THE FRUSTRATION OF WORKING FOR OTHERS

Are you an engineering career victim? Are you ready for a promotion that never seems to materialize? Are others in your company passing you in salary, or position, or in their ability to get more interesting projects? Have you been shuffled off to the company sidelines because you are too specialized? Do you work in constant fear of being laid off? Do you suffer from the negative thinking and "can't do" attitudes of your co-workers? Are you bored or underutilized in your work? Do you want more say about the way things get done? And most important, are you getting the most mileage out of your knowledge and abilities? Chapter 1 shows you how to leave all these problems behind by becoming a consultant!

CASH IN ON YOUR UNIQUE SKILLS!

Have you worked long and hard to attain your technical skills? You deserve to cash in on them! In Chapter 2, I tell you why you *must* assume responsibility for your own advancement and explain why it is so hard to do that within someone else's company. You will learn why technical abilities are only *half* the story and why you must learn certain business skills if you want professional independence.

I show you why you never have to worry about competition if you develop the ability to *find* situations that match your skills. I tell you how I beat the big-time consulting companies by finding a project that matched my skills exactly! And in Chapter 15, I show you how to *develop* new areas that expand your capabilities.

MARKETING SECRETS MY BOSS NEVER TOLD ME

Want to make your customers eager to obtain your services? It's not easy, but it can be done by learning how to be an expert in something they just happen to want. You're not selling oranges, and the way you create demand is different from the techniques used by your neighborhood grocer. The financial rewards are spectacular for those who learn the tricks. In Chapters 4, 5, and 6, you learn the real story behind the marketing of technical consulting.

DEALING WITH CLIENTS IS A SNAP—IF...

you know what you're doing. This book gives you a crash course on client relations. You learn the techniques for putting both your customer and yourself at ease. I tell you about my "one-over" method of dressing for client visits, and give you my list of things *not to be done*—*"nevers"* that could save your professional rear end! Finally, in Chapter 11, I show you how to negotiate with the client, how to deal with unrealistic client expectations, and how to build the kind of trust and customer satisfaction that keeps 'em coming back for more of what you have to offer.

THIS BOOK TELLS YOU HOW TO GET STARTED FROM GROUND ZERO

Many business advisers have limited vision. They can tell you about something only if it *already* exists. Well, chances are that you are not *already* a consultant, and you're wondering what the first steps are. This book gives you sufficiently detailed guidance to make a smooth transition to consulting. For example, in Chapter 7, you'll learn how to sharpen your sales abilities. I'll show you my "secret weapon" that makes clients salivate with interest! The financial aspects of starting a practice are concisely explained in Chapter 12. I'll tell you why a business plan is vitally important in starting your company. You'll get an idea of typical start-up costs for office and equipment and learn the tradeoffs in locating an office in your home versus commercially rented space. In Chapter 14, I'll help you write a brochure for your new business that will display your "wares" and credentials to maximum benefit.

YOU CAN CREATE YOUR OWN FUTURE

If you are stuck in a go-nowhere job, it may seem hard to believe that you have *any* control over your future. But with some encouragement and self-understanding, you can set your goals and "engineer" your own future! In Chapter 15, I'll tell you about my experiences with goal setting. In becoming a consultant, you are not pursuing a *single* business goal, but a whole hierarchy of them. What's so special about this way of describing goals? It shows very clearly *which* goal is the one you must tackle first! And it shows why you can't help but reach your highest professional goals if you follow the planning and evaluation guidelines.

HOW MUCH ARE YOU WORTH?

Many engineers and technical people go through training every bit as intensive and demanding as doctors and lawyers, but they can't seem to make their unique background pay off. Instead, they stick with low-paying positions in large firms and feel sorry for themselves. It doesn't *have* to be that way! You can earn in excess of $100 per hour by selling your specialized technical abilities!

I remember being told by a well-intentioned boss many years ago, "We think your performance this past year has been spectacular. But we can't give you a raise, because money is tight this year and because you are already at the top of the salary schedule for your level..." When you become your own business entity, your worth is not limited by a salary schedule or any other factor that is unrelated to your actual contribution. As a consultant, you can also pyramid your wealth by expanding into a multiperson company—your company—or by collecting lucrative commissions for deals that *you* put together. Your income is limited only by the scope of your entrepreneurial vision and initiative!

PROPOSALS AND CONTRACTS

Afraid of writing proposals? Don't be! They are your *key* to winning high income, and they are your chance to show that you can do the job better than your competitors.

In Chapter 8, I explain the basics of proposal writing for the technical consultant. You learn what to say, how to phrase it, and how to price your efforts. I illustrate the subject with real-life proposals, contracts, and quotation letters to give you a concrete idea of the formats. Things are much easier when you have straightforward examples to follow!

A BOLD STATEMENT ABOUT THE REWARDS OF CONSULTING

Are the rewards of consulting worth the effort? Let me answer the question this way: In all my years of engineering, I have *never* met a person who was happy working for another company after being his own boss! Never! Once a person has tasted professional independence, experienced the monetary benefits, and felt the power of determining his own future, he will never go back to a lesser situation!

IS CONSULTING FOR YOU?

I think consulting is absolutely the best career path available to engineers and technical specialists, but it may not be everybody's answer to professional satisfaction. For some, the needed combination of talents, credentials, and attitudes is simply not there, or is too difficult to obtain. For others, the desire to be a part of a larger business organization may be more important than professional independence and high income. To find out if consulting is for *you,* turn to the unique self-appraisal quiz given in Chapter 17.

Harvey Kaye

ACKNOWLEDGMENTS

I wish to acknowledge the many people who have helped me in my consulting career and given me the encouragement to explore new ideas and areas.

A Zen master is said to inspire not by teaching, but by setting a living example. With respect to engineering consulting, my father, Joseph Kaye, was such an inspiration. He showed me what was possible, and I acknowledge his great influence.

I also wish to thank Jim Esselstyn and Kathleen Wheaton for encouragement and moral support during the writing of this book.

I am indebted to John D. Constance and Eugene D. Veilleux for reading the manuscript and making helpful comments and suggestions. Particular thanks are due to Leonard M. Schwab, who reviewed the manuscript in detail and contributed a number of refinements and clarifications. Also, I wish to express my gratitude to the "Meet the Pros" individuals for sharing their experiences in the sidebar pages of this third edition.

My sincere appreciation goes to Frank Cerra, my editor at John Wiley & Sons for the first two editions. Thank you for your editorial support and guidance in getting this project off the ground! Finally, I would like to thank Robert Argentieri and Minna Panfili at John Wiley & Sons for continued encouragement and support on this third edition.

H. K.

CHAPTER 1

How to Leave Your Career Problems Behind

ARE YOU AN ENGINEERING CAREER VICTIM?

The rank and file of almost every engineering firm is full of engineering career victims. Some of these engineers suffer from inadequate pay or low status in their company. Others complain of being treated as a hired hand and not as a professional. Some engineers feel that innovation and professional development in their companies are not only unrewarded, but actually discouraged. Many competent engineers also find progress very difficult beyond a certain point. The individual may be brilliant in his particular area and become known as the "heat transfer person" or "computer architecture expert" in his company. But he can't seem to break through to new areas and responsibilities for which he may be ready. The discouraged contend that the really clever engineer will get out of engineering per se and go into management. As one cynic put it, "Engineering is a profession where one must cease to be an engineer in order to become successful."

This victim psychology will not help the individual grow toward a mature understanding of his career. As the newly graduated engineer starts out, there is no explicit promise that the employer will be committed to optimizing the engineer's advancement. In this respect, many engineers have been confused by unrealistic expectations fostered by engineering schools and corporate personnel offices. The plain fact is: A corporation does not owe its employees much. The engineer, as well as most other corporate employees, is indeed a hired hand. But the engineer's ego will not allow him to sense this business relationship. It is too painful to consider that he could be replaced in the same way as a floor sweeper. The engineer develops a certain blindness, therefore, to the business realities surrounding his employment conditions.

THE SECRET OF SUCCESS:
DEVELOPING "GROWTH VISION"

A person must learn to develop the perspective that will allow him to grow from where he is *psychologically.* Yes, engineering does have many negative aspects and shortcomings as a business area. But this is not to say that a person cannot become successful in engineering. As in many areas of life, success is possible if time and energy are expended to build up personal confidence and enthusiasm, *despite* the many shortcomings of the field itself. The price of not becoming a victim is hard work and learning to see the positive goal and not the ocean of negatives surrounding it.

Why am I starting a book on technical consulting with a discussion of psychology? Because most engineers work in an environment that is not conducive to developing attitudes of professional independence or business self-identity. After years of marching in line, many engineers become hypnotized into "can't do" thinking. They focus on reasons things won't work or how cost, time, and expediency make change unfeasible. After prolonged exposure to this environment, they absorb their organization's inertia, inefficiency, and confusion. They assume that they also must behave this way.

CLAIM CREDIT FOR YOUR INNATE CREATIVITY

Innate creativity is destroyed by being ground down by the system of established, bureaucratic engineering organizations. This is not an accusation leveled against organizations. For one is not "ground down" unless one willingly accepts that fate. After ten or twenty years on the job, many engineers develop a slave mentality. They see no evil, hear no evil, speak no evil. They have become *less.* They are just trying to coast until their retirement day. These people are sometimes referred to as "company men" and "company women." Do not look to such people for advice or encouragement, as they do not value the very qualities you will need to survive in consulting: creativity, efficiency, and professionalism.

On many engineering projects, the basic design is dictated by the top manager, regardless of its feasibility. The details are left to the engineering staff, and they are sometimes stuck with the difficult task of making an unfit design "work." I can recall an example of this situation that occurred in a prestigious international firm. The boss invented an all-in-one machine that performed the function of three separate machines. The only problem was that it couldn't be maintained or repaired at reasonable cost. Each person under this manager who was given the task of implementing the invention came to the same conclusion. However, the price for telling the manager "no way" was a demotion, as three successive subordinates were to find out.

DIRECT EMPLOYMENT VS. CONSULTING— A FUNDAMENTAL CHANGE

As a direct employee, you can become extremely frustrated by not being able to benefit from your own knowledge and judgment. As a consultant, the situation changes in a fundamental way. If you are in a situation where you are not being effective or feel that you are compromised, you have the freedom to find new work at your convenience, without the onus of job-hopping.

Please don't think that consulting is a bed of roses! On the surface, it might appear that consulting is an activity in which

- you don't have to answer to anybody;
- you can do anything you want;
- there are unlimited funds to pursue professional interests;
- you can now do "creative" things;
- all business expenses are "free."

This is not the case. Moreover, the consultant will find a new and different set of constraints that now apply to him as a small business entity. However, for some engineers, consulting may be the best way to maximize professional goals and personal independence.

Operating your own consulting practice has a number of significant advantages over direct employment in a large company. Your small business has the flexibility to react to changing markets and business conditions. Large corporations have so much inertia that they may not be able to change rapidly enough to take advantage of new situations. Capital equipment considerations may even lock a large company into a waning or unprofitable market.

The consultant can refuse assignments that she knows to be potentially hazardous to her credibility. In contrast, a large company may allow an individual to be a sacrificial lamb in situations where the company's welfare is at stake. Of course, the direct employee has the option of refusing a particular task, but that usually results in the ultimate loss of employment.

GAIN CONTROL OVER YOUR FUTURE!

One of the greatest advantages of consulting is that you have much more control over the direction of your career. Much of the engineering work done in large companies leads nowhere in terms of professional advancement or growth. If you are a corporate employee, it is difficult for you to control what projects you will be engaged in, and you may not have the vision to understand where each particular project leads. Once you are assigned to a long-term project, it is usually impossible to extract yourself from it without leaving the company.

Life is too short to allow yourself an extended series of projects that lead nowhere! Consulting tends to minimize the impact of these situations and allows you more control over your work content and conditions. You are free to pursue projects that will advance your career in the strategic direction that *you* choose.

Some people argue that since a decision to become a consultant is reversed simply by reentering the ranks of direct employment, the decision does not warrant great introspection. I do not agree. Such a major change should be approached with intelligence and deliberation. Credibility, and ultimately, success—either as a consultant or as a direct employee—depend on the ability to make clear strategic decisions and hold steady to them. This book attempts to give prospective consultants sufficient details to determine the suitability of consulting to their own situation.

THE CHANGING FACE OF CORPORATE AMERICA

A number of significant trends are emerging in corporate America—trends that spell difficulty for directly employed technical specialists and, at the same time, opportunity for self-employed consultants. These changes are making it much more difficult to find a rewarding position as a direct employee. The notion of working for a single company for ten or twenty years is gone forever.

In the 1950s and 1960s, the demand for engineers and technical people was so strong that "jobs came looking for the engineer." It was a seller's market. When an engineer graduated from college, he or she would get ten or twenty employment offers without any effort on his or her part. Scientists and engineers graduating with a master's degree were wined and dined at the finest restaurants by prospective employers.

The demand for technical help was so strong that nearly everyone prospered and advanced. Even marginally competent direct employees rose up the technical ladder to become high-level managers. Consultants had an easy time, too. They could approach industrial companies or government agencies with a two-page proposal and get contracts worth hundreds of

---

**The Changing Relationship between Company
and Employee**

- Less loyalty — both ways
- More insecurity
- Fewer benefits
- Longer hours
- Less time for professional development
- Less patience
- Less trust, more litigation
- Fewer defined paths to professional advancement
- More competition for the few top positions

---

thousands of dollars. (I watched my father, who was consulting at that time, do exactly this!)

All this has gradually changed in the last thirty years. The supply of qualified technical help has increased faster than demand. The number of new engineers and technical workers entering the workforce has increased dramatically since 1970. In the eighties and nineties, new engineering, scientific, and computer graduates flooded into the job market. Colleges and universities pushed technical degrees as guaranteed "meal tickets" in the workplace. They failed to warn students that jobs might not be waiting for them when they graduated. Of course, the universities were in the middle of their *own* crisis: financial survival mandated keeping enrollments high to offset the severe cost inflation that plagued the higher-education market. The universities needed to attract enough students to fill their classrooms; it was not in their interest to disclose the faltering employment prospects that their graduates might face.

The sixties were the pinnacle of opportunity for technical specialists. President Kennedy created a national priority to put a man on the moon by the end of the decade. The space program created a sudden demand for thousands of engineers. Yet, as Apollo and the other space missions were completed in the early seventies, the demand for engineers and scientists saw an equally sudden decline. The layoffs were massive. Many engineers complained that job opportunities were impossible to find.

In the early eighties, political pressure to reduce the national deficit forced Congress to cut back on defense spending. The defense cuts were mild at first, but by the late eighties, with the end of the cold war in sight, the defense budget was *slashed*. Defense R & D was hit particularly hard. The defense industry, which in the seventies had fueled about one-third of the total demand for technical labor in this country, was now greatly oversupplied with technical specialists. Opportunities slowed down even more. Engineers lamented how hard it was to find decent work in new industries.

In the nineties, reductions in the corporate workforce and cost-saving pressures reduced the demand for technical specialists even further. The fancy word for this effect — *disintermediation* — does not assuage the pain of its ax. To control costs and consolidate efforts, many companies chopped middle management and senior technical positions. The result was a loss of experienced workers and reduced quality. "Management by the bottom line" was starting to bankrupt technical innovation in America's largest companies.

Yet another factor in the supply-demand relationship was the emergence of global sources of supply. The United States, once the premier producer of technology specialists in the world, experienced a gradual erosion of its leadership in this field. By 1985, Japan, Great Britain, Germany, and many other countries were producing more technical graduates per capita than the United States. Many of these foreign technologists were every bit as capable as the best the United States could offer.

Moreover, thanks to the Internet and new telecommunications technologies, working from a remote location is no longer a great liability in many lines of technical work. In computer programming, for example, jobs are being shifted offshore. An entire project is given to a software development firm in Calcutta. The Indian programmers know the latest programming languages, speak perfect English, and are extremely industrious. They post daily progress reports to their client in Chicago via the Internet. Most important, the offshore programmers cost one-half to one-third what their American counterparts cost!

The final trend affecting the demand for technical specialists is a lack of national policy about technology goals. Current governmental tax policy provides few incentives for corporations to take risks with new technical developments. As a result, R & D efforts, which use large numbers of technical specialists, are put on the back burner.

The United States government has been reluctant to initiate new priorities itself, even when they are desperately needed. For example, over the past thirty years, each administration has promised a national energy policy but has failed to deliver. This problem transcends partisan politics. Congress establishes large and costly bureaucracies to "investigate" and "recommend solutions," but nothing happens. Similar patterns have emerged in transportation, medical care, food technology, space exploration, and home-building technology. I firmly believe that technology is the only way to solve certain of the important problems facing us in the twenty-first century. In energy policy, for example, we simply do not have sufficient oil reserves to continue our present wasteful methods of home heating and automobile propulsion. We need viable alternatives that are more efficient. Such alternatives will come through technological initiatives that are, in part, accomplished by technical consultants.

In previous editions of this book, I referred to the reduction in demand as an engineering *recession.* By now it is clear that *recession* is the wrong word to describe what has been happening. Recession implies a cyclic nature, which has not been the case. The trends have pointed down for a long time, and they are still headed down! This is not a recession in the classical sense, but a subtle shift in the economy at large. Author William Bridges calls this shift the "de-jobbing" of America. In his book, *JobShift,* he comments,

> It isn't your fault that you were brought up on a diet of Get-a-Good-Job-and-You're-All-Set. Of course you bought into it. Everyone did....It isn't your fault that you were born just in time to get swept up in the Second Great Job Shift [the first being the Industrial Revolution]. Everyone is in the same boat today.

Bridges suggests that the fundamental way American business is being conducted is changing radically. Companies that use technical help of any kind—direct or by consultants—are changing their traditional modes of operation. To stay alive in today's new market, they must now move faster, stay leaner, and strive toward higher performance. More important, they must offer better *value* to customers. Because competition has intensified, successful companies are now more strategically focused, which means that they want their projects done faster, better, and more knowledgeably.

There are many ways companies are accomplishing these objectives, among them *reengineering* and *downsizing.* Michael Hammer and James Champy coined the word *reengineering,* which they define as

> the fundamental rethinking and radical redesign of business systems to achieve dramatic improvements in critical, contemporary measures of performance, such as cost, quality, service, and speed.

Hammer and Champy's book, *Reengineering the Corporation,* hit the top of the best-seller lists in 1994. Subtitled *A Manifesto for Business Revolution,* it created a sensation in the business world by offering faltering companies a method for streamlining business operations and increasing profitability. At its best, the book stimulated a cultural change that has encouraged many companies to adapt to the new market realities. At its worst, the book created false hopes that there was some simple panacea that could save poorly managed companies from disaster. For those companies, *reengineering* became a euphemism for laying off employees, outsourcing, and cutting benefits.

The fallout of these structural changes is only beginning to register on the public at large. Companies and stockholders may be happy about the improved bottom line, but employees are bitter and wary. Having a staff of loyal and long-term employees is no longer an important company goal. The vagaries of the marketplace force companies to regard employees as temporary and replaceable. It is often easier to use a temp, hire a consultant, or outsource entire company functions and projects rather than build up a costly employee structure. As a result, employee loyalty evaporates. Workers must work harder to make up for the people who are no longer there. They are told that if they don't work longer hours for "free," they will be replaced by individuals who will. Cynicism skyrockets as companies

- lay off older workers who approach pension vesting;
- ignore bright young workers who clamor for professional advancement;
- refuse to grant workers reasonable annual salary reviews;
- reassign, bump, or demote workers arbitrarily;
- tell workers they care — and then cut their benefits;
- ask workers to perform duties for which they are not trained;
- eliminate workers as the result of a power struggle between divisions or a merger of divisions.

Although large companies become leaner and more efficient this way, customer service often gets worse. This is obvious in the software business, for example, where customer support is no longer free, even if the difficulty you are experiencing is attributable to a bug in the program design. Similarly, customer service at banks has become more "streamlined," which means that they will no longer count a bottle of loose change for you with their coin-counting machines.

The people hit hardest by downsizing are in the middle of the management and technical ladders. Their ten- or fifteen-year investment has just gone down the tubes. There is increased competition among the remaining managers, so their chances of getting promoted are even lower than before. Downsizing has a way of dead-ending everyone except the few individuals at the very top.

With reengineering, employees are being asked to work faster, learn on their own time, and put in longer hours, often as much as seventy hours per week. Direct employees have few good defenses against being asked to work these long hours. "Whipping the ponies" is the company's way of getting more from the employee so that it can remain competitive in the marketplace.

The current situation is in great contrast to my first job at the Avco Everett Research Lab. In 1967 technical specialists were still scarce. The company wined and dined me on my interview. Once I was working there, the company sent me to numerous technical conferences and picked up the tab for my professional journals. I was given an attentive annual review during which my boss discussed how the company would help me advance professionally. Very few companies treat their employees the same way today.

One side effect of this new economic order heralds a very rosy future for consultants. It is the dramatic growth in temporary labor firms. Once restricted to secretarial and clerical areas, the "temp" industry has expanded to technical specialists as well. Its growth has been nothing less than spectacular. The country's largest employer is not IBM, General Motors, or General Electric, but Manpower! The age of contract technical work has arrived, and consultants are in a great position to take advantage of it.

Consulting doesn't eliminate the problems discussed here, but it allows you to charge for each hour you work, to plan for your learning ("product improvement"), and to avoid layoffs when your projects are completed. If you are underutilized, undervalued, or in constant fear of being laid off in your present job situation, perhaps you should consider consulting. If you are *already* laid off, or have found unsatisfactory reemployment after being forced out of your last company, consulting may offer an excellent way of achieving your career goals.

BENEFIT FROM MY EXPERIENCE

I started this book in 1976, when I was in the process of becoming an independent consultant myself. I searched for background reading on how to start a consulting practice. The search was frustrating. Many books described how to start a retail business and how to merchandise products, but few books applied directly to engineering, scientific, or computer software consulting. Information from business associates and friends often turned out to be inaccurate or colored by pessimistic thinking and personal prejudices. "Success" and "get-rich" books told me that I could achieve *anything* I wanted, if only I would "believe." But these books did not give sufficient concrete information for me to make reasonable evaluations. Nowhere was I able to find a simple description of the steps involved or the factors important to success.

I ended up learning the techniques of consulting by starting my own practice. It was not simple. Before I made the transition to full-time consulting in 1976, I spent two years in planning and preliminary activities. I cannot say that consulting has made me very rich (wealth being the usual measure of success in our society). But I have earned significantly more than most directly employed engineers. And I feel very good about the stimulation, challenge, professionalism, and freedom that consulting has given me.

On a personal note, certain aspects of my background made consulting a natural activity for me. My father, Joseph Kaye, was a well-known M.I.T. professor and consultant. The memory of his enthusiasm and energy has always been an inspiration to me. I had the credibility of three M.I.T. degrees and a list of publications to my name. The desire to be professionally independent was a high priority for me. And I had already worked as a direct employee of two large consulting firms, where I gained familiarity with technical problem solving and client relations.

This book is written in the hope that you may gain from my experiences and grow in your own career.

CHAPTER 2

Defining Your Own Business Identity

Farmer Smith, of Lexington, Virginia, bought a prize rooster at the state fair for the highest price ever paid in the history of the poultry trade. When he got it home, he discovered, to his chagrin, that he could not control the rooster's romantic inclinations. Not only the hens, but the ducks, swans, and geese, as well as a few stray nanny goats, fled from the rooster's tireless onslaughts.

Farmer Smith finally could stand it no longer. Collaring the overzealous bird, he complained, "I did not pay a record price for you to waste your energies on every form of animal life in Virginia. From now on, you are to confine your activities strictly to the hens. If you continue with these shenanigans, you'll die of exhaustion."

The rooster made light of the farmer's admonition, but sure enough, a few mornings later, Smith found him flat on his back, with his eyes glazed and his legs straight up in the air. A couple of buzzards ominously circled closer and closer above him.

"What did I tell you, you blazing fool?" roared the farmer. "I knew that this crazy life you were leading would kill you sooner or later!"

But then, to his amazement, the supposedly expired rooster opened one eye and whispered, "Pipe down, will you? When you're romancing buzzards, you've got to play it their way!"

One of the most subtle dangers in our complex and compartmentalized society is that we play at so many roles and identities to get ahead that we forget who we really are. The matter of business identity is just one aspect of this problematic situation. There have been many times when I have felt like Farmer Smith's rooster, trying to "score" by assuming business identities that weren't mine. What I have discovered is that I didn't really need to *identify* with some other business entity's goals and values, but that I could claim my own.

CONSULTING: THE FAST TRACK TO SUCCESS

Many people feel that if a sufficient number of "nice" direct positions were available on the job market, there would be no need for consultants. Talented engineers would naturally be attracted to large *firms,* and each firm would be able to meet its own needs for specialized technical help. This common attitude appears reasonable, but it neglects a number of fundamental business facts.

People often go into business for themselves because they want to make faster progress than is generally possible in a large corporation. Large companies hire a great number of engineers and technical specialists, and it makes sense that only a few can achieve significant and rapid advancement. Many engineers in the system are happy making slow progress because that situation is consistent with their aims in life. In fact, most technology firms are geared to provide jobs for the mediocre. If your aim is to put in extra effort to advance quickly, you are bucking the system.

HOW THE SYSTEM "REWARDS" DIRECTLY EMPLOYED PROFESSIONALS

For the directly employed engineer in a large corporation, the financial rewards are *not* directly proportional to individual performance. The system undercompensates the overachievers and overcompensates the underachievers. The most talented and productive engineer I know is vice president of research for a medium-size company. Although he has some freedom to pursue technical areas of his choice, his financial compensation is only 50 percent higher than that of the average engineer. Thus, a three-sigma effort gets a one-sigma increase in salary! Likewise, the least competent engineer I know always manages to find work after he has been laid off. These layoffs are the "let's clean out the deadwood" variety, and he usually gets caught in them. But he finds another job in a few months, and his overall earnings are perhaps only 40 percent less than that of the average engineer. Thus, on the negative side, a three-sigma individual suffers only a one-sigma salary penalty.

Working for a large corporation is an *advantage* if you are "average" and is a *blessing* if you are below "average"!

Many engineers of moderate talents make it to the top of their companies simply by putting in their years. Talent and technical performance are not the sole criteria of advancement. Other factors such as effort, patience, loyalty, and the ability to deal with people can carry more weight. Nevertheless, if you are aiming for a top position in a large company's engineering section, it is worth considering the small numerical probability of your rapid advancement.

THE MAGIC OF DOING YOUR OWN THING

Advancement and financial reward are not the only reasons for becoming a consultant. The individual may want to specialize in a field that is of great interest to her, but not central to her employer's needs. For example, an engineer may want to specialize in advanced numerical methods to analyze gas-dynamic shock waves, an area in which she may have done extensive research or written many papers. Her company needs such work occasionally, but not regularly. Becoming a consultant allows that individual to seek other clients who also have an occasional need. From the aggregate, the consultant creates a "position" in which she is working in the area of her foremost technical interest.

One of the reasons you are probably reading this book is that you are above average and are wondering how to maximize your situation. Many engineers and technical specialists feel the same way, yet there are only a limited number of desirable positions within established companies. The individuals waiting in line in front of you may have spent considerable effort in dedicated service and in forming political and social alliances within their companies. With such a large supply of technical talent in this country, specialists can be treated as *expendables* rather than as *resources*. That is, the technical specialist can be hired or laid off as the project work load dictates. Reacting to this fact emotionally serves no purpose, for it will only blind you to the objectivity needed to make good career decisions.

Technical brilliance is becoming commonplace as more and more engineers enter the marketplace. The market premium for this characteristic alone is decreasing as years go by. As a consequence, marketing yourself will become increasingly important in the future. In this respect, having your own business identity can be a great asset, for you will never forget *who* is responsible for doing the marketing.

My father, Joseph Kaye, provided me with a firsthand image of an engineer who took responsibility for his own marketing and business identity. In addition to teaching at M.I.T., he conducted an active consulting practice where he experienced the magic of doing his own thing. The consulting brought him into contact with many of the best minds in the country. The resulting intellectual and professional stimulation had a dynamic and energizing effect on him. Further, he had the satisfaction of earning high income using his special talents.

For many engineers, the fruits of fine creative talent are often limited by their company's inability or unwillingness to utilize their individual talents. Not so for my father. He did not wait to be "assigned" a project; he was an idea generator who developed his ideas for the joy of it (and the money). He was in the ideal situation where his output was not limited by his employer, but only by his own energy and creativity!

YOU ARE RESPONSIBLE FOR YOUR OWN FUTURE

I have come across many talented and well-intentioned engineers who have not lived up to their professional potential. Their own negative thoughts have kept them in a rut. They lose genuine interest in their career and begin to live for outside interests such as bowling, tennis, building a new porch, or buying a new boat. These outside activities are certainly not negative in themselves, but they should not be substitutes for the excitement, challenge, and fulfillment that can be derived from professional work.

Be careful not to give up and rationalize a poor attitude. Giving up is the "minimum energy" option, but it is not an appropriate reaction to the lack of incentive offered by your company. It is a mistake to *assume* that a corporation—or any other organization representing the interests of its owners—will "love" you for better or worse. *No professional should assume that some other business entity will look after his or her own long-term welfare!* You cannot relegate this responsibility without paying the consequences.

I do not mean to suggest that an individual working for a large company should be disloyal or selfish. However, you should continuously monitor whether your career goals are being achieved within that company. Other opportunities available to you may lead you closer to your goals.

YOUR UNIQUE KNOWLEDGE MAKES YOU VALUABLE

Why is it possible to make a good living from technical consulting? We all see advertisements to "earn millions raising rabbits in your cellar" and to "become financially independent selling high-protein supplements to your friends and neighbors." *Anyone* can do these activities. Your technical degrees and professional experience make you a scarcer commodity. The key to success in the engineering consulting business is to find the opportunities and profit from them. And to learn to *keep* your "commodity" up-to-date and marketable.

A TRUE STORY: HOW A SMALL CONSULTANT BEAT OUT THE BIG GUYS IN A COMPETITIVE BID

The fact that every technical specialist has a unique "collection" of experience and skills is one of the reasons a consultant never has to worry about competition from other consultants. As an example of this principle, a few years ago, I was asked to bid competitively on a contract for technical support in the following areas:

- Heat exchanger design
- Optical distortion effects
- Laser thermal blooming
- Turbulent flow theory
- Scale model testing
- Gas distribution manifolds
- Mathematical modeling of flow networks

The requirement was for a single individual acquainted with *all* of these areas. Although there were twenty bidders, I was able to win the contract. It happened that the requirements looked like a replay of my résumé, and I was able to take advantage of my unique background. (A few prestigious consulting firms proposed *teams* to address the technical scope, but their hourly rates were much higher, and they were dismissed in the client's selection process.) The point is that there are contract possibilities where *almost every* experienced engineer has a unique advantage over her competition.

VISUALIZING YOUR OWN OPTIONS

When we start out in the workplace, most of us have little capital, scanty business experience, and even less *vision* of how we would like to grow professionally. Most of us have been taught from an early age that we will work for a company — someone else's company. Technical training in college and graduate school does little to change this. We are offered no formal instruction in entrepreneurship. It is in the colleges' best interest to convince us that *technical* courses are the key to success in the corporate world.

Since many people have difficulty visualizing anything other than direct employment, I present some creative visualization techniques here. Indeed, from one perspective, this entire book can be considered a visualization guide: As the different aspects of consulting are described in the following chapters, imagine yourself in those situations. Paint *yourself* into the picture!

Tip

In *Secrets of Self-Employment,* Sarah and Paul Edwards describe twelve mental shifts required to make a successful transition from employee to independent business owner. They emphasize that self-employment is not "business as usual."

> We expected life to work the way it had when we were employed. We thought if we followed directions carefully, we would succeed. But sooner or later, operating from a paycheck mentality can make being self-employed a confusing, frightening, and pressure-filled experience.

The Edwards' book is an excellent guide to the new set of freedoms, benefits, *and* responsibilities that self-employment entails. It is highly recommended reading.

Are you happy being there? Do you feel professionally empowered in that situation? Does being there as a consultant feel right to you? Used this way, this book offers a glimpse into another possibility, so that you may begin to adjust your career plans to your own long-term advantage.

Visualization helps in making significant career transitions because such transitions involve issues of *identity*. We can't "decide" to change our identity overnight. Identity is our concept of who we are, and it changes only gradually, as we try on new hats and see how they suit us. We look in the mirror, ask others how the new hat looks, or wear it for a day to see how it feels. Sometimes simple "yes-no" answers don't emerge from these tests right away. Yet, one day, we wake up and realize, "Wow, I've left that old hat called 'direct employment' behind. I'm the boss of my own company now!"

Identity shifts are subtle and elusive psychological processes. When I started consulting, I was only 32 years old and had been a direct employee for nine years. I didn't even *think* of myself as a consultant until I had been doing it for two years! In retrospect, I realize that this situation is normal. First, we dare to dream of something new for ourselves. Second, we begin to conceive of how we might reach that dream. If our evaluation looks promising, we take it to the next level by declaring a *goal* to ourselves. Next we act on that goal, which, in effect, declares it to the world as well. Only after we do it for a few years will others begin to acknowledge our new role of consultant. Finally, we assimilate these experiences and acknowledgments into our identity; they become who we are.

Visualization involves letting go of some of our perceptual biases and seeing what new options are there for us. This "letting go" is not as easy as it sounds, for most of us have a heavy psychological investment in maintaining our current situations—even if we hate them! Our current situation,

---

MEET THE PROS UP CLOSE AND PERSONAL

Larry King is the founder of International Information Services, a Virginia-based firm specializing in system engineering and computer consulting. Larry received his bachelor's degree in mathematics in the sixties and a master's degree in MIS in the seventies. Over the  years, he has held a variety of project management positions for the defense industry. One day in 1994, Larry's boss invited him into his office and described how his projects were being phased out. Larry's services were no longer needed. In a word, Larry was being laid off. At age 55, Larry found himself with two weeks' severance pay and dim prospects of finding a good direct employment situation.

Larry is a positive thinker who believes that "difficulties are the breeding ground for opportunities. How you perceive your situation is up to you. Attitude is the key." Although things didn't look great when he left his employer, Larry took stock and realized that he was in a position to capitalize on his experience and contacts by becoming a consultant.

Starting with short notice and limited financial resources, Larry found consulting to be a struggle at first. He reflects, "It was a good thing my wife had a steady job as a teacher. Learning how to market myself and my services was a great challenge." Within a few months, though, Larry developed leads from family and friends into paying contracts. His big break came in 1996, when he was simultaneously approached by a client in need and by two people who wanted to work for his company. Larry now has three full-time employees working for him. Relationships with his employees have proved rewarding because Larry operates by the Golden Rule, not by bottom-line management.

Larry says that direct employment offers little security these days. It's hard to build a track record when the track gets changed every few years. Consulting, on the other hand, has been very satisfying. "I'm the only one who can fire me. Now I have some control over my destiny."

however distasteful it may be, is still a known quantity, whereas an unknown choice carries with it the risk of dangers we cannot yet appreciate. We feel locked in by our past choices, powerless to change without assuming great risk.

Therefore, the first preliminary is to affirm your willingness to act on your own vision, to affirm that you are resourceful enough to make things happen, and to affirm that you deserve success. The second preliminary is to make room for the new to appear. Create the openness to see what's really out there. To do this, distance yourself from your ordinary psychological surroundings, which, in this case, means your everyday fears, worries, and doubts. The third preliminary is to temporarily drop all labels and identities that you already have. Allow yourself to float free for a moment. Affirm your freedom to change into situations that bring you greater professional satisfaction.

Once these preliminaries are done, just relax and pay attention to the patterns that emerge. These patterns can take many forms—for example, trends, opportunities, or situations that seem to beckon to you. Then, imagine yourself in those situations. See how they look, feel, taste, and smell. This "emotional taste-testing" is like painting a picture. Make it as complete and detailed as you can. Capture every nuance, every new angle. You don't need to be a Picasso to do this kind of painting—we all have this creativity built in! As a result, you will develop a set of mental images of yourself in that new role, hopefully enjoying it and succeeding in it.

LEARNING SURVIVAL SKILLS

In our free-enterprise society, individuals have the right to choose their form of business identity. Individuals can choose to be sole proprietors, partners, corporation owners, or employees of other firms. There is more to the story than *choosing,* though. *Survival* is the other half of the story. And, just like in the jungle, the survivors are the ones who are most fit.

The consultant should consciously recognize that professional survival depends on being "fit." In this context, fitness means more than technical expertise. It includes learning essential business skills and how to implement them in practice. The consultant should know *where* to find business, *how* to conduct the business, and *when* to adapt to changing environments. The consultant must understand not only her technical specialty, but also the way the "jungle" works—or else she will eventually fall into the quicksand or the tiger's mouth!

This book will help you learn these survival skills. Chapters 4 through 6, which deal with marketing, show you where to find business and how to nurture contacts with clients. Chapters 7 through 14 survey the essential business techniques you will need as a consultant. And Chapters 15 through 19 teach you (forgive me, dear reader, for milking this analogy) how to understand the jungle tom-toms to determine your next moves.

CHAPTER 3

Who Becomes a Consultant?

HOW MUCH EXPERIENCE IS NEEDED?

There is no definite answer to the question of how much experience is required to be *consistently* successful at consulting. A relatively inexperienced person can win a consulting contract and competently perform the work. But it is one thing to do this on an isolated basis and another thing to do it time after time.

Experienced consultants know that even a few contracts in a row do not guarantee that a practice will continue to be viable. Long-term success depends on being able to *plan* and work toward predictable results. You must learn *how* and *why* you got your contracts if you want to re-create your victories consistently.

By learning to objectively understand your customers' reasons for giving you contracts, you will develop the ability to be more effective in your marketing efforts. It is important to know whether you won on the basis of lowest price, immediate availability, your unique abilities, or any other major factors. It is crucial to appreciate whether you are filling a need for expertise or supplying additional labor to satisfy a temporary peak work load. If you don't have this knowledge, the results of your efforts will he haphazard and you will spend much of your time chasing the wrong contracts and the wrong clients. These marketing aspects are discussed further in Chapters 4 through 6.

TECHNICAL ABILITIES ARE ONLY HALF THE STORY

Generally speaking, engineers and technical specialists in their twenties may have the *technical* ability to consult, but they probably lack the business experience and energy needed to start an enterprise. At this age, most technical specialists are enjoying their newfound prosperity in the workplace and are perhaps establishing families. A great amount of energy is spent on acquiring a home, taking care of infants, and so on.

By the time most technical specialists reach their early thirties, their personal lives have settled somewhat and more energy can be devoted to career issues. The experience and insight needed to make good business decisions are usually attained by this age. Part of this insight is the understanding that a consulting practice is a major commitment that requires a reshifting of personal priorities.

I recommend *not* going directly from school (even graduate school) into consulting. It's far better to start out in a large company, where you can gradually learn the required business elements, understand how the industrial system works, build a set of worthwhile credentials, make professional contacts, and hone your abilities as a technical problem solver. Working as a direct employee at first also allows you to accumulate the start-up capital to finance your consulting practice. Finally, without some industrial experience under your belt, you may be unable to appreciate *why* consulting is a desirable alternative to direct employment. You'll come across Scott Adams's *Dilbert* cartoons and wonder if the corporate workplace could really be that full of idiot bureaucrats, idiot bosses, idiot policies, and idiot co-workers.

REAL-LIFE CASES: CONSULTANTS WHO MADE IT AND THOSE WHO DIDN'T

To give you an idea of the variety of situations in which technical specialists enter (or leave) consulting, consider the following examples. All names have been changed to protect the privacy of the individuals described.

1. John was the manager of computer-aided design for a medium-size consulting company. After fifteen years on the job, he knew the technical angles very well and felt he had sufficient contacts to establish his own consulting practice. He found a client who was willing to give him a sizable contract once he was officially established. John put up very little cash initially, and in three months he hired three junior associates to do the work while he did more marketing.

2. Owen is one of the country's leading experts in infrared optical system design. After twenty years of working for the industry giant in this field, he became a consultant. At 40 percent utilization, he earns enough to pay his bills. He uses the rest of his time to manufacture energy-efficient furnaces as a side line. He is a happy man.

3. Yappi is the nickname of a famous M.I.T. professor who "wrote the book" in his technical area. He is both brilliant and practical. He makes consulting commitments of one-day duration only, which he schedules between class days. More often than not, he solves a problem in a few hours, but if he does

not know the answer, he is not afraid to say, "I don't know." He is well worth the $2,000 per day he charges his clients. With his stellar reputation, he does not have to write proposals. His business deal is simple: If you want me to come and look at your problem, it will cost you $2,000 for one day. Period.

4. Ron started consulting in his late twenties with a bachelor's degree, eight years of experience, and no professional publications to his credit. He had difficulty obtaining work of an interesting nature and spent a few years working as a contract engineer. One of his clients noticed his talents and offered him a direct position as a project manager. He is relatively happy in this capacity.

5. Susan is a civil engineer who is fluent in Spanish and familiar with Spanish culture. In college, Susan attended the University of Madrid for a year on an exchange program. She liked it so much that she spent the following three summers in Spain, working for an architectural firm. Following graduation, Susan went to work for an engineering firm in the United States. Within ten years, she had responsibility for interfacing with three major Spanish client companies. Susan developed such rapport with these clients that she decided to become a consultant servicing their needs. Specifically, she is a "go-between" who helps Spaniards and Americans write and negotiate bilingual technical contracts and proposals. Business is booming and will continue to do so, since the trend toward globalization is increasing.

6. Bill was the mechanical engineering manager of a small design company. His skills lay more in the social area than in engineering. He wined, dined, and entertained his customers in a grand fashion. A close buddy gave him a million-dollar contract, with which Bill started his own consulting firm. He hired capable associates to do the work and spent lavishly on offices and attractive support personnel. But his flamboyant attitude and expensive tastes were out of line with engineering work, and he went out of business as fast as he went in.

7. Don is a sales genius. After fifteen years as an engineer, he became marketing manager for his employer. He knows the technical angles very well, but he has no interest in being tied to a desk or in doing detailed calculations. He has contemplated going into consulting for himself many times, but each time, he has decided against it. For him, this has probably been a good decision. In his present position, he can delegate many of the less desirable chores, and he has the freedom to pursue sales as he sees fit.

8. A group of eight distinguished scientists decided to leave the large corporation for which they worked. These people all knew and trusted each other and had a common goal of business self-identity. To avoid conflict-of-interest problems, they started out with no contracts in hand. But they had the maturity and experience to know that they had a solid market and a good reputation in their field. For the first year of operation, business was slow

(as planned), and they maintained a modest overhead. In their second year, sales boomed. Because they built on quality, they are doing very well today.

9. Murray is a likable guy who is an expert in a special analytical procedure that is in great demand in the power industry. He has twenty-five years' experience, degrees in engineering and business, and a professional engineers license. In the three-year period preceding his transition to consulting, Murray worked for a large firm, marketing and performing his specialty. One day he decided he had enough contacts to make it on his own. He started out with no contracts in hand, but he had assurances from his five best contacts that they would place him on the bidders' list when he left his employer. He is now very successful.

10. "Whiz" was a Harvard M.B.A. graduate with eight years of technical management experience. He left a plush position with a prestigious firm because he felt he should be advancing faster. Without really thinking through his plans, he decided to become a computer software consultant since that was "where the action was." He had few real contacts in this field and found much of the work mundane. "Whiz" felt he had no prestige as a consultant compared with his previous high-status, high-profile position. He deplored the long hours and uncertainties. One day a headhunter approached him about a good management position and he accepted without a second thought.

11. Herman is an electrical engineer by training. With his flair for writing and marketing, though, he soon found himself writing technical proposals for his employer. Herman discovered that there was a ready market for helping other companies with their proposals, so he became a technical proposal consultant. He no longer does electrical engineering but teaches people how to structure and write winning proposals. Herman's lucrative practice now consists of writing books, giving seminars, and consulting on how to write and market technical proposals.

12. Jim is an expert in computer methods applied to a very specialized area of electrical engineering. He studies the latest literature and goes to meetings where industry leaders discuss their needs. He then goes back to his home office and creates computer programs that address these needs. He markets his products to companies that pay a handsome royalty to use his programs. He is also retained by a number of his clients to provide user support for his software. Jim is making a great income doing exactly what he loves.

13. Helen and Mary went to engineering school together and became close friends. Helen was the quiet, introspective analyst, while Mary was the extrovert who became class president. Her technical skills were not as outstanding as Helen's, but she related well to people and found it easy to sell herself. Ten years after graduation, they compared notes and decided to leave their unfulfilling corporate positions and start a consulting company together. In this successful partnership, Mary was the marketer and Helen the technical guru.

14. Phil is a statistician by training. After fifteen years at four different companies, he wound up in the quality control section of a large firm. The company was repressive in its policies and work environment, so Phil decided to strike out on his own. He determined that there was a ready market for quality audits at companies too small to have their own quality department. He was so successful in filling this niche that he eventually hired two employees of his own to help with his growing client load.

CONSULTANTS GRAVITATE TO "NICHES"

It may seem difficult to generalize from the above examples about personal characteristics and situations that are conducive to success in consulting. Consultants gravitate to niches in which they have a natural advantage, and therefore, many characteristics are purely individual. Nevertheless, the following attributes appear to be common:

- A solid grasp of their technical specialty
- Knowledge of the industries involved
- Desire for professional independence
- Strong communication skills
- Resourcefulness
- Ability to deal with people

Why is finding niches critical to consultants? Part of the answer comes from the way technical work is procured in this country. Companies no longer want to train their own technical specialists. It is too costly, and there are too many uncertainties. Why invest three years of company money to "breed" a specialist, when the specialty may be obsolete by the time the training is complete? Most companies resolve the problem by trying to hire individuals on a direct employment basis. That is why so many technical employment ads are frustratingly specific. The company has a very specific need, and it won't spend one dime or delay the schedule by one hour to train someone.

At the same time, a large part of America's technical workforce has been unable to keep up with technological advances. These technical workers are definitely not stupid; they have been sidetracked into repetitive work and managerial tasks, or are burned out and no longer want to stay current. Many engineers now in their forties and fifties were burned out ten years ago and have been coasting since! They are technologically obsolete, but the company needs them. These "elders" are the only ones who remember company history, policy, and procedures.

Even though the client may have a dozen specialists in your technical area, those specialists may still be too "general" to handle a project that requires specific skills and knowledge at the cutting edge of technology. Try as the client may, finding a direct employee familiar with the cutting-edge technology is often impossible. This is where the consultant comes in. Consultants zero in on these hot niches of opportunity. They become better, faster, and smarter than everyone else—in that very narrow field. By pinpointing their technical efforts, they offer the client greater value in terms of technical expertise and cost-effectiveness.

TECHNICAL CONSULTING IS DIFFERENT FROM MANAGEMENT CONSULTING

Some people feel that all consultants can be conveniently lumped into a single group (the "rose is a rose is a rose" philosophy). For the moment, however, I wish to distinguish between technical consultants and *management consultants*. Technical consultants must possess a commanding grasp of their technical specialty. Strength in technical issues gives them a competitive edge and reduces the need for highly developed business and "people" skills.

Management consultants cater to a different clientele and require a different set of skills. In most cases, management consultants market the top managers in the front office. This requires a more polished corporate image and finely honed people skills. You need finesse and diplomacy to play the "front-office game." You must be much more articulate to coax reluctant CEOs into accepting your managerial vision. Most management consulting deals with projects that are people and decision oriented. Success in this area depends more on judgment and experience than on rational methods of analysis and design.

Engineering consultants, on the other hand, tend to market their clients' engineering managers, who give little attention to boardroom etiquette. All they want is to find credible ways to solve specific technical problems.

Table 1 compares the skills you need to succeed in technical consulting, management consulting, corporate management, and other positions. In considering your options, see if these skills are well matched to your choice. For example, suppose you have modest technical abilities but strong business and marketing skills. Maybe you should become an entrepreneur instead of a consultant. Similarly, if you have outstanding business, marketing, and people skills, consider becoming a management consultant. Finally, if you have well-rounded skills but are not a technical "heavy hitter," you might be better suited for corporate management.

TABLE 1. Skills Required for Various Positions

Position	Business	Technical	Marketing	People	Presentation
Business Entrepreneur	High	Low	High	Medium	Low
Technical Consultant	Medium	High	Medium	Medium	High
Corporate Manager	High	Medium	Low	High	Medium
Corporate Technical Employee	Low	Medium	Low	Low	Low
Professor: Part-Time Consulting	Low	High	Low	Low	Medium
Management Consultant	High	Low	High	High	High

FIND YOUR OWN "STYLE"

Beyond the common characteristics mentioned above, I find a wide variety of personal *styles* among successful consultants. It is important to understand what your style is and how to go about getting things done your way.

Don, the super-salesman mentioned above, loves to "sell" consulting but hates to do the detailed technical work. He now concentrates on the marketing, and his staff of technical experts performs the work.

Neil did poorly in the first few years of his consulting. He was trying to make it as an individual consultant, when, in fact, he wanted to be a team player. Being part of a team was an important and rewarding experience for him. This showed in other areas of his life as well. He was a member of the local orchestra and contributed actively to the local chapters of his professional societies. He was able, over a period of time, to figure out that he would also like to be part of a technical team. He found five other consultants who had professional skills that complemented his and organized a team. This team now bids as a unit on projects that would be unobtainable to each individual. But more important, Neil has that feeling of being "part of the band," which is central to his satisfaction.

"Pops" is a consultant in his early fifties who has a total mastery of his technical specialty. He has eight children and truly enjoys being a father. In his personal life, he leads a Boy Scout troop and talks enthusiastically about their shared educational and constructive projects. It is no wonder that he gravitates to consulting situations in which he is working with younger engineers. In his last consulting project, he was the technical expert for a group of recently graduated engineers who had just joined his client's company.

Whatever your personal style, awareness of it will lead to consulting possibilities that are truly satisfying.

YOU DON'T HAVE TO BE A GENIUS

A popular misconception about consultants is that they must be geniuses or internationally known experts. This is simply not true. A person with sufficient qualifications will find his or her own level within the consulting community. As a word of encouragement in this regard, it is my observation that the great majority of consulting work done in industry requires merely competence and not brilliance.

This book *presumes* that you, as a prospective consultant, do, in fact, have a basic competence in your specialty. If you do not, don't despair, but consider how you may gain such competence in the next few years.

If you ask certain successful consultants who received their Ph.D.'s from big-name engineering schools what qualifications you need to establish yourself in consulting, they may project their own history onto you. They will say that you need the same academic credentials. Further, they may inform you that, if you are not brilliant, clients can, and will, tear you to shreds in technical meetings. These messages enhance their image in front of you, but, at the same time, scare you away from trying.

What are the facts behind the issue of qualifications? Most engineering consultants do *not* have their Ph.D's. Almost all have a bachelor's degree, and many have master's degrees. A few have no college degree at all. (But you can be sure that they have other compensating qualities!) Although many consultants have graduated from prestigious universities, most have degrees from a wide spectrum of engineering schools. Therefore, do not worry whether your academic background "proves" you intelligent enough to enter consulting. Focus instead on developing a set of salable skills at which you are competent!

Real intelligence is the ability to make your life into what you want it to be. *Anyone* can do this, whether or not you have three degrees from M.I.T. And, conversely, M.I.T. degrees are absolutely no guarantee that a person has the *real* intelligence to make it happen!

THE STORY OF DAVID AND GOLIATH, RETOLD

Qualifications and credentials are not the only factors in consulting success. You don't have to be the country's foremost authority in your specialty to get interesting projects. How do I know this? Well, let me tell you about my David and Goliath experience.

A local power company had a pressing technical problem that forced its plant to sit idle until a fix could be found. A "tiger team" was assembled to solve the problem. One of the team members was an acquaintance of mine who realized that the problem involved one of my special areas, fluid transients. He asked me to come to the kickoff meeting.

At this preliminary meeting, the project managers and task members were all introduced. Lo and behold, one of the invited participants was the country's most prestigious name in fluid transients, whom I shall call John Smith. He was a professor at a well-known university and chairperson of a society committee on fluid transients. He had been invited by the project manager, who had called his friends at M.I.T. and asked for the most respected name in fluid transients in the country. (It turns out that this famous professor was charging about three times per hour what I was charging.)

The famous authority listened attentively at the meeting but did not make a single comment or ask a single question. At the end of the meeting, the project manager decided to go around the table for comments, suggestions, and action items. When it came to Professor Smith's turn, he had to say something, for the problem was right down his alley.

Professor Smith thought the problem was "interesting"; he could use it as a Ph.D. thesis project for one of his graduate students. Here was a client with an extremely urgent problem, and this expert was suggesting a solution in three years! The manager asked Professor Smith if he could generate some *approximate* numbers that would suggest a fix. After all, the power company didn't want to advance the state of the art in fluid transients, it just wanted to solve its problem. The professor offered that he would go back to the university, think about the kinds of calculations he might do, and send back a proposal in two weeks.

It was my turn. I outlined my thoughts about the significant issues and parameters. Then, in general terms, I showed how a few simplifications would allow the problem to be solved *approximately*. I proposed a method to determine whether the simplifications were justifiable. And finally, I said that I was fascinated by the problem, and that I could give them an answer (not a proposal!) in four weeks.

My knees were shaking. Would the famous professor attack my suggestions? I had seen him turn uncomfortably in his chair while I was at the blackboard sketching my ideas. He hadn't said anything technical the whole meeting, and here I was, saying exactly how I would do the problem. Would he put on the dark green glasses that he used for doctoral candidate qualifying examinations and ask me to derive my assumptions from basic Navier Stokes theory? Would this Goliath heave a huge stone at me? No. He didn't say a word.

It turns out that the professor never delivered. He went back to his university and thought about the proposal. The problem was not "clean" in the sense of a controllable scientific experiment. He didn't want to furnish the client with an answer that wasn't 100 percent accurate. He couldn't verify his assumptions. Part of the project was outside his narrow area of expertise, and he was too busy to bone up on it.

The professor must have felt that, *for him,* the risks of the project far outweighed the potential benefits. *Not* responding was his best course of action. I gave the customer what it needed and obtained its thanks and three follow-up contracts!

DESIRE TO HELP THE CLIENT CAN GO A LONG WAY

Professor Smith was the embodiment of the *mandarin consultant.* This is the guru who has long fingernails and does no work. He merely nods "yes" or shakes his head "no." Even better, he is totally noncommittal and says, "Ah so!" to everything.

Even though Professor Smith had credentials that far exceeded mine, he was of no *service* to the client. He viewed the meeting as a means of soliciting a three-year funded research program. He had no intention of providing timely answers. He was more interested in furthering his own causes than those of the client. This attitude is common among big-name experts[1] and is another reason a small consultant can beat an industry giant. David won the contract because he was *sufficiently* qualified to do the job *and* because he wanted to be of service.

CONSULTANTS AREN'T ALL MAVERICKS

Although most consultants are ethical, a few are not. Sometimes, an unscrupulous consultant tries to scare a client into a course of action that benefits *only* the consultant. Sometimes, maverick consultants make insincere promises in order to land a job. When they can't deliver, they create elaborate technical excuses for not delivering. And at other times, wayward consultants manipulate clients into buying equipment and services that are totally unnecessary.

As in all ethical breaches, there is a price to be paid. When the client finally realizes that he's been had, that's the last time he'll ever do business with that consultant. Unfortunately, a sucker is born every minute, and so unethical consultants just move on to their next victim. As long as they don't do anything overtly illegal, they can continue to weasel their way out of responsibility and ply their tricks on unwary clients.

---

[1] It's ironic, but the more stellar the consultant's reputation, the lower the client's chances of getting a job done satisfactorily. Big-name experts can afford to play a different game than most other consultants.

For this reason, clients usually ask for references from previous clients if they don't already know you personally. Once a client has been stung, he is much more careful. Clients who are stung many times eventually develop a cynical attitude toward consultants. They feel that *all* consultants are mavericks and not to be trusted. Consultants seem to be free spirits and not bound by corporate allegiances. But, in fact, most consultants *are* respectful of their clients' organizations. Respect and being of genuine service are an important part of ethical consulting. Just remember that a mature and ethical consultant does not confuse respect for the client with the loyalty expected of employees toward their corporation.

CHAPTER 4

Three Fundamental Truths of Technical Marketing

MARKETING YOUR SKILLS: A LEARNING EXPERIENCE

Strong skills in marketing are essential to the consultant's long-term success. These marketing skills are not attained in a short time. Years of practice are necessary before a consultant feels comfortable meeting new clients, writing proposals, and negotiating with purchasing agents. It takes experience to discern the key factors in a business situation and recognize the critical aspects beyond one's immediate knowledge that must be researched before proper responses can be made. A limited business background should not deter a qualified person from consulting, but those new to the marketing and business areas should make allowance for learning and a few fumbles.

FIRST FUNDAMENTAL TRUTH OF MARKETING

I feel there are three fundamental truths about marketing technical consulting services. The first truth is: *The need for consultants is genuine.* There are thousands of engineering, science, and computer consultants in this country. Most of them earn a very high income and do useful and interesting work. Their success depends on a number of factors, but above all, having specialized knowledge that is *salable.*

Many engineering companies normally operate in an understaffed mode. When a sudden increase in work load or an emergency arises, it may make more sense to hire a consultant than to commit additional permanent staff. In other situations, a company may have an occasional technical problem that is so specialized that nobody within the company is qualified to handle it. Considering the urgency of the need, hiring an expert may be the best way to effectively address the issue.

SECOND FUNDAMENTAL TRUTH OF MARKETING

The second fundamental truth of marketing is: *Getting the work is more difficult than doing the work.* This is true of engineering in general as well as consulting in particular. The competition for projects is intense. This situation is neither good nor bad. It is simply how things are. Individuals working within large corporations may not be conscious of the free-enterprise aspects of their existence. However, consultants are acutely aware of what "free enterprise" means in terms of marketing their own services.

Graduate schools are notorious for conditioning students into believing that their technical training is their most valuable asset. This may be true of the technical specialist employed by a large firm. But for the consultant, the ability to survive depends just as much on *marketing* as on specialized technical skills.

WHY YOU NEVER HAVE TO WORRY ABOUT COMPETITION

Competition is part of the free-enterprise system. Other consultants and certain of the client's direct employees could do the same specialized work that you do, although perhaps not as efficiently or elegantly. Overcoming this competition is achieved by having a good "commodity" *and* by bringing this commodity to the marketplace. Those people who have the goods but not the means to bring it to the marketplace will not survive in consulting for long.

From a business point of view, marketing is the number one priority: if it is lacking, all else is in vain. There are many successful marketers who first "sell" a job and then hire the specialized help they need to do the job. On the other hand, those who are totally lacking in marketing skills or interest in developing these skills must learn to be good corporate (or government bureau) citizens. They must suffer the prisonlike aspects of being told what to do, how to do it, and what will happen to them if they don't do it.

If competition is a fact of life in technical consulting, why am I insisting that you won't have to worry about it? Because you'll be able to overcome the competition with repeatable results when you follow the marketing strategies and procedures in this book. As you will see shortly, the answer is to *position* your consulting uniquely and in a way that clients find irresistible. When you do this, it will be easy to get your foot in the client's door while other consultants are waiting outside.

Even if you are relatively young or inexperienced in your technical field, there are strategies you can use to maximize your competitive position. For example, in fast-moving areas of technology, there are no "established" experts (that is, people with twenty years of experience). Not as much experience is required to consult in these new fields. If you are considerably younger than the average consultant, then head into a rapidly developing field. Stay away from mature fields in which an army of established experts have been plying their trade for many years; their extra experience and credibility with clients will prove unbeatable.

Competitive strategy bears a striking resemblance to military strategy. In fact, Karl von Clausewitz's classic book on military strategy, *On War,* provides a decent introduction to marketing strategy as well. Clausewitz wrote his treatise in 1832, long before the invention of the machine gun, the military jet fighter, the guided missile, and the command-and-control telecommunications center. Yet, the *principles* of competitive strategy remain the same. They are as relevant today as they were hundreds of years ago. You still must figure out:

- What competitive goals are attainable, considering my resources and strength?
- Where do I attack? Head on, at my strongest spot, or at competition's weakest spot?
- Which actions will bring the best results?
- *When* should I attack? Now, or wait for an opportune moment, such as when the competition falters or exposes a major weakness?
- How do I concentrate my forces to break through a defense?

 Tip

In *Marketing Warfare,* Al Reis and Jack Trout provide an excellent introduction to Clausewitz's strategy as applied to marketing. For example, the three principles of offense are:

1. The main consideration is the strength of the leader's position.
2. Find a weakness in the leader's strength and attack at that point.
3. Launch the attack on as narrow a front as possible.

The authors give a lively discussion of how offensive, defensive, and flanking strategies are used in marketing today and illustrate Clausewitz's concepts with vivid examples from modern business.

Once you understand the principles of strategy, it's easier to adapt your actions and plans to account for changes in the overall market, changes in the competitive landscape, and changes in your own situation. By considering the classic elements of

- Timing
- Strength
- Gathering intelligence
- Positioning
- Tactics to outwit opponents
- Deploying a campaign

you will become more successful in your technical marketing, regardless of whether you are defending, attacking, or flanking the competition.

A consultant starting out is the strategic equivalent of a small country with few resources except determination, flexibility, and little to lose. You can afford to be more daring and innovative than large consulting companies. Moreover, since established individual consultants have an enormous investment in their present course of action, they tend not to digress widely from their fields of expertise. Instead of competing with you in new areas, they will spend their energy maintaining a significant market lead in their present area. This strategy is described by Clausewitz as "defending a stronghold."

WHAT ARE YOU SELLING?

The consulting services you offer in your technical specialty become your *product*. Viewing the service as a product is useful in a discussion of marketing because it forces you to define your marketing approach.

If you have a narrow specialty, achieving a satisfactory business level involves finding those scattered few customers needing this service. The marketing approach is to locate and identify the needles in the haystack. If there is sufficient market for this specialty, it may be worthwhile concentrating in it and developing a competitive edge by furthering the state of the art. The liability with this approach is that the specialty may become obsolete and your competitive edge may be useless in a field where the demand is waning. This will be discussed in greater detail in the last section of this chapter.

Although many consultants start their careers in a narrow specialty, they ultimately expand into related areas in which they have developed some capability. While this expansion of capabilities is desirable from a marketing viewpoint, it is important not to spread oneself too thin technically. Therefore, the key to the expansion of capabilities is that they should be in related areas and done in a *planned manner.*

As a consultant, you should not be afraid of venturing into new areas just because you aren't the world's foremost expert in them! What counts is the degree of expertise *relative to your client.* That is, your credentials in a rapidly growing technical area may not be as convincing as in your field of expertise, but you may still know significantly more than your clients. And, in fact, you may be their best bet to accomplish the work at hand.

There are two general liabilities in accepting work outside of your main specialty. First, in consulting in an area of relative inexperience, you can blunder. Making mistakes is not terrible. Everyone makes mistakes. But *blunders* are mistakes of the variety where everyone feels that you should have known better. Making too many blunders can destroy your credibility. The antidote to this situation is to openly explain to your client that you're on a "best efforts" basis for work that is outside your area of expertise. (If the customer has been pleased with your work in the past, there is usually no difficulty with this arrangement.) Second, if the work is totally unrelated to your specialty, it can detract from your main line of business. After a while, your customers—and you—will become confused about what you are selling.

YOUR MARKETING "MIX"

Every client company has its own set of particular technical needs that a consultant can satisfy. Although you may think of yourself as a heat transfer specialist, your clients will think of you in even more specific terms. Company A may need you to design cooling systems for its large computer modules. Company B may need you to analyze thermal deformations in its missile O-Rings. Company C may need you to review its heat-pipe design. To Company A, you will become the module cooling consultant. To Company B, you will become the O-Ring thermal deformation expert. And at Company C, you will become known as the heat-pipe guru.

The concept of *marketing mix* refers to the fact that you must present a slightly different "face" to each and every client. To me, this is a good thing because it means I won't be doing the exact same analyses over and over. A reasonable amount of variety keeps me vital and interested. To think about your marketing mix, concentrate on what your clients' needs are and how your clients define individual tasks in your general area.

For example, while designing cooling systems at Company A, you may notice that the client has difficulty performing a thermal reliability analysis for those same systems. Perhaps you could expand into this area with a few weekends of self-guided study and offer a new service that is closely related to what you are already doing. Such a capability might be well received, since you're already familiar with the client's hardware.

Every consultant has a bag of tricks—areas of competence that they offer to clients. The questions are: How many areas must you master? How many different tricks do you need? How do you expand your current bag in a way that new tricks are synergistic with the old ones and continue to give you a natural business advantage? The answers lie in matching your clients' various needs to your ongoing personal development efforts.

The marketing "mix" of services you offer should reflect your particular interests and background, as well as your situation in the marketplace. Become aware of the general approach and specific strategies you are emphasizing in your own practice. To aid this process, a few questions for self-examination will reveal much about your marketing approach:

- What is your main product line (area of expertise)?
- What are your *technical* subareas within that area?
- Which industries have need of your specialty?
- What services are required by each industry?
- Which of your technical subareas will appeal to the industry segments you have targeted?

To assure a sufficiently wide marketing mix, try to address a half-dozen industries in the first five years of your consulting practice. If you put all your marketing eggs in one industry, when that sector has a temporary downturn, you'll be scrambling to establish a *new* customer base. However, when your market mix is sufficiently broad, the chances of all market sectors turning down at the same time are minimal.

To better visualize your marketing mix, draw a matrix with your technical subareas running down the side and the industries you are addressing across the top. For consultants in individual practice, this matrix is usually no greater than six by six. By checking which of your "tricks" are used in each industry, you can gauge whether your mix is too narrow or too broad. In Table 2, a sample matrix is given for a heat transfer specialist.

IMPORTANCE OF ACQUIRING BROAD BUSINESS KNOWLEDGE

I urge you to acquire a broad knowledge of technology and business for the purpose of *communicating* with the client's staff. You will be interfacing with managers, accountants, designers, and engineers in other fields. If you can't

TABLE 2. Marketing Mix for a Heat Transfer Specialist

Technical Subarea	Industry				
	Aerospace	*Computer*	*Petro chemical*	*Electric Power*	*Home Appliance*
Fan cooling systems	Y	Y	Y	Y	Y
Computer simulation	Y	Y		Y	Y
Conduction cooling	Y	Y			Y
Heat pipe design	Y	Y	Y		
Cryogenic heat transfer	Y				
Aero shock waves	Y				

follow the client's discussion of her overhead rates, her instrumentation problems, power consumption, etc., you are in a poor position to understand the context of her total problem. Here, "follow" does not mean to be an expert, but to have sufficient acquaintance to listen intelligently and ask reasonable questions. Managers tend to discount the recommendations of consultants who they feel are too parochial to appreciate the cost, manpower, schedule, legal, manufacturing, and reliability aspects of the project.

THE THIRD FUNDAMENTAL TRUTH

A basic concept of marketing is that every product has a *life cycle*. That is, every product experiences a trajectory that starts at the R & D stage, progresses to early market conditions, peaks, and then fades as new products offer more benefit or higher performance. It is important to understand that the same process occurs with bodies of specialized knowledge.

If you are learning your consulting techniques from a college textbook, the chances are that your area is already obsolete with respect to consulting opportunities! You must find *new* areas within your field as you go along, or you will be chasing dwindling markets. The third fundamental truth is: *Engineering thrives on the novel.*

As a recent example of this product life cycle in consulting, consider the history of finite-element stress analysis after its introduction in the early 1960s. When this useful and powerful technique for analyzing complex structures emerged, only a few professors and consultants knew the complicated theory and computer algorithms. These consultants enjoyed an active market until the mid-1970s. By then, numerous college textbooks had been written to explain the procedures in detail, and many of the proprietary computer codes were supplanted by publicly available codes. Every new mechanical engineering graduate knew about the technique from his coursework. This field became less fertile for the consultant. Those consulting companies that had developed lucrative finite-element stress analysis services became aware of changing market conditions by the early 1980s. Many were able to successfully expand their consulting into related areas. Finite-element stress analysis remains an active and useful tool in the engineering world. But it is no longer a very attractive field for *consulting*.

HOW TO GET IN THE "NOVELTY GROOVE"

If novelty is a major driving force behind technical consulting, how can you make sure that your "product line" addresses this requirement?

I can't offer a surefire technique, but I can point you in the right direction with a correlation I have noticed over the years: The boundary zone between *two* emerging technologies offers the most fertile ground for consulting. That is, it is not enough to master *one* emerging technical area; you must master *two* to create the conditions for successful consulting. The cross-fertilization between two distinct areas or competencies almost always provides a mix distinctive enough to attract clients. This concept will be explained more fully in Chapter 6, where I discuss creating your unique market niche. But for now I present just the principle: Look for the boundary zone between two emerging technologies!

For example, Steve is a C++ programming expert who spent a lot of his own time looking into bar-code-reading technology. His original motivation was to create a system to track and inventory his personal collection of compact discs. He researched all the manufacturers of bar-code equipment and knew the trade-offs between the different devices on the market. When a local client advertised for a programmer to develop a bar-code inventory system, Steve brought his CD system into the client's office and showed how it worked. Two weeks later Steve landed the contract. Although many other knowledgeable programming consultants responded to the ad, none was able to catch the client's attention with such a convincing demonstration.

THE BEST CONSULTING LIES ON THE GROUND FLOOR

The desirability of consulting work on a particular project can be related to the location of the project in its life cycle. (See Figure 1.)

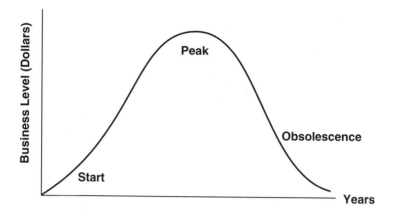

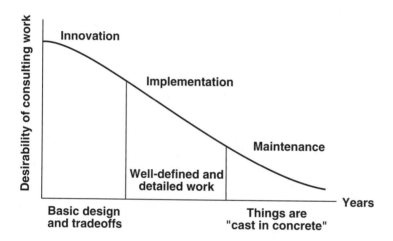

FIGURE 1. Product life cycle.

When a project or product starts out, creative people are usually required to conceive and investigate the initial stages. These people are often the most competent in the company, and the atmosphere surrounding this endeavor is very stimulating. All the key engineering trade-offs are done in this phase. The demand for consultants is highest here, because the technical issues are fluid and unexplored.

As the project matures, the preliminary design is completed. The remaining work requires practical experience and implementation more than insight. The tasks are much better defined, and the dollar volume of expenditure is high while the details are being executed. The opportunities for consulting are not as exciting, for the consultant faces two obstacles. First, the project already has a history, so the people on the project know more than the

MEET THE PROS UP CLOSE AND PERSONAL

Jim Lacy is the president of Jim Lacy Consulting in Dallas. He is the author of *Systems Engineering Management: Achieving Total Quality,* published by McGraw-Hill. Trained as an electrical engineer, Jim has a P.E. license and twenty years of experience gained at  Texas Instruments, where he created a number of groundbreaking courses in systems engineering, project management, and quality improvement. He is also a founding member of the International Council on System Engineering (INCOSE).

Jim says that consultants must learn to narrow their focus to become more successful. In 1983, when Jim first experimented with consulting, he offered general electrical engineering services. He discovered that he could only command "journeyman's rates," $35 per hour. To improve his situation, Jim started to focus on a narrower area: systems engineering. The basics were already in place, so all Jim had to do was create credentials in that specialty and find a way to get the public's eye. He wrote a unique series of courses called "Systems Engineering for Total Quality" and presented them to hundreds of people. As his seminar experience deepened, he had enough material to write a book on the subject. In addition, Jim started going to IEEE professional society meetings. At first, he joined committees as a participant. Soon, though, he volunteered to write some new standards. Jim offered to do extra work that the other committee members were too busy to do, and within one year, he became chair of the EIA committee. At the end of a three-year period, Jim was no longer a journeyman, but an expert in his field.

Jim's advice to beginning consultants is "Find something you like, and then build credentials in it. Don't be one of the many, but one of the few." Following his own advice has paid off handsomely. Since 1992 his charge rate has been $1,975 per day (nearly $250 per hour). And customers are willing to pay his price because he offers a unique combination of skills and credentials that indeed make him "one of the few."

consultant about the particulars. As a result, the consultant has less influence and must work hard to catch up with the preliminary design rationale and results. Second, the nature of the work is more detailed and offers less exposure to the management staff. This is a disadvantage from a marketing viewpoint because the consultant has less opportunity to demonstrate his or her insight to the client's decision makers.

Finally, the project nears completion, and the company's low fliers, the "grunts," are brought in. One sure sign that this stage has been reached is when the accounting office appears to be dictating all the important decisions! By this time, all the bright engineers have left the project to work on more attractive ones. It seems to take forever to wade through the pile of documentation required to understand what the system does. Very few people know why things were done the way they were: "Jones did it ten years ago, and he has since left." At this point, the design is cast in concrete. Trying to suggest better ways to accomplish the goal will fall on deaf ears. The client wants only to complete the project in the most expeditious way. Consulting in this situation needs patience and skill, and is not likely to lead to new business.

CHAPTER 5

Marketing Secrets Other Consultants Won't Tell You

THE SECRET OF FINDING CLIENTS

At times, clients will be seeking you and work will seem to just fall into your lap. That situation will not last forever. Sooner or later, you will be faced with the question of how to generate new clients.

Everyone who becomes a consultant wonders at one time or another: What is the secret of finding clients? Secret? If anything, the secret is that *marketing involves more work than anyone would ever guess.* Consequently, it is imperative to make your marketing efficient, to systematize it.

The marketing system you ultimately adopt will depend on the nature of your specialty, the particular industries you service, market conditions, and your personality. Developing your own system requires the willingness to experiment and see what works. Trial and error is necessary because there are so many variables. "Copycat" marketing will simply waste your precious resources. Blind imitation in marketing—as in other areas of life—has dangerous consequences:

> It was a busy day in the courtroom. Three women had been arrested for soliciting, and a peddler, who had recently arrived in this country, had been arrested for peddling without a license. They were brought before the judge.
>
> "What do you do for a living?" the judge asked the first woman.
>
> "Your honor, I am a model," she replied.
>
> "Thirty days" was the sentence. Next the judge turned to the second woman. "What do you do for a living?" he asked angrily.
>
> "Your honor, I am an actress."
>
> "Humpf! Thirty days." The judge then confronted the third woman. "What do you do for a living?" he demanded.
>
> "To tell you the truth, I am a prostitute," she answered.
>
> The judge blinked. "For telling the truth," he said, "I am going to suspend your

sentence." Finally, the judge turned to the immigrant peddler. "And what, pray tell, do you do for a living?"

"To tell you the truth," the peddler said, twisting his worn cap in his hands, "I'm a prostitute also."

To make the marketing function easier to grasp, I'll outline the general procedures for you here. Of course, you'll have to modify them to suit the particular nature of your own practice. For technical consultants, marketing is a seven-step process:

1. Have something to sell.
2. Become known as a credible source.
3. Find out who your customers are.
4. Make contact with your customers.
5. Get your foot in the door.
6. Develop a marketing plan.
7. Set your plan into ACTION!

Please keep in mind that I am ordering these steps for convenience in exposition. In practice, marketing involves juggling all these activities at the same time. In this chapter, we shall look at steps 1 through 5 in detail. Chapter 6 will cover steps 6 and 7.

HAVE SOMETHING TO SELL

The first step in marketing is *defining your service.* This is not as trivial as it sounds. You must decide not only what you are offering, but also what you are *not* offering. Further, what you are offering must be *salable.* That is, there must be market demand for the service you offer, and you must be able to persuade clients that they should pick *you* rather than one of your competitors. Define your service to stand out from the crowd. Offer something unique.

For example, suppose you are a heat transfer expert with degrees in mechanical engineering. You have unique accomplishments and credentials as a heat transfer expert. Yet, heat transfer is just one small area within the general field of mechanical engineering. Does that mean you must forget about all those courses you took in mechanics, materials, thermodynamics, and manufacturing? Yes and no: Yes, you should forget about them in terms of defining your consulting specialty. Hit your clients with your best shot.

No, you should not forget the useful techniques that you learned in these courses and the rest of your technical curriculum—they provide the breadth you'll need to make sound technical judgments and to communicate with technical specialists in other fields.

Mechanical engineering has become such a vast field that you can't be an *expert at large* in it. You must pick the subarea in which you have the most strength. That is, market yourself as a heat transfer expert. Don't even bother to sell yourself as a mechanical engineer—you won't be able to distinguish yourself from the pack.

With the explosive growth in technology recently, even the subarea of heat transfer may be too general to differentiate yourself in the market. If you try to market yourself as an expert in conduction, convection, and radiation, you will probably meet with little success. These three modes of heat transfer cover all the bases, but they're still too general to attract the attention of clients. Instead, consider how the clients in your target market segments think of your services. To the computer market segment you might sell "computer cooling system design." That's how *they* think. They want someone who can cool their computer, not someone who knows everything about convection. To the air-conditioning market segment, you might sell "high-performance heat exchanger design." And to the appliance industry, you might sell "thermal testing for UL® qualification."

Whatever discipline you studied in school, the principle remains the same: Market yourself in the *most specific* terms possible! That is, create a niche (or a few niches) in which you are unique.

ADVERTISING: THE ART OF CREATING DEMAND

Many engineers and scientists who have been immersed in a purely technical environment find it difficult to understand the need for advertising. To some, the word *advertising* connotes unethical or subprofessional behavior. If you feel this way, it is worth remembering that the consultant does not have the marketing and sales departments of a large firm to handle these aspects for him. He must shoulder the responsibility himself.

For consultants, advertising does not take the same form as in the merchandising field. In fact, most consultants find that placing advertisements in local papers and engineering journals is a poor way to develop new business. Listing in the yellow pages is also ineffective, because it presumes that your customers will be seeking you. In most cases, *you will be seeking your customers,* so a different kind of advertising is called for.

Indirect advertising is appropriate for the engineering consultant. In indirect advertising, the consultant gains the attention of potential new clients by coming into their view as a technical expert. Success in indirect advertising means *becoming known as a credible source of technical expertise.* There is no offer to sell services, no sales pitch. The natural forums for accomplishing this are:

- Making presentations at professional society meetings
- Teaching professional short courses
- Participating on society technical committees
- Writing technical articles
- Authoring books in your field
- Publishing computer software in your specialty

If you are writing an article, interviewing potential clients for their opinions on the article's subject matter is an excellent way to build contacts. You get to meet the person in a purely technical context, without the pressure of a sales meeting. Be sure that you have first written a few articles to which you can refer. You will need these prior articles to legitimize the capacity in which you are acting.

For example, a few years ago I wrote a paper on the relationship between integrated circuit junction temperatures and system electronic reliability. After putting together most of the material, I asked a number of reliability managers in local firms about a certain technical issue in the paper. These people were happy to grant me time, as these issues were of great concern to them, and because my interview was about a very *specific* question. Of the ten people I spoke to, I developed a natural rapport with about five. After the paper was published, I sent a copy to these five people, and three contacts developed from it.

In making a presentation to a professional society, the consultant has the opportunity to demonstrate himself as an authority in his field. Good presentations have a lasting effect on the individuals in the audience. How many presentations should you aim to give each year? I am tempted to answer, "Between two and five." But for marketing purposes, it is not sufficient to merely "give" a few presentations every year. Quality, not quantity, is what counts. The consultant must prepare adequately to ensure a smooth presentation with accurate information and relaxed delivery.

I find that people's response to technical presentations is geared more to the delivery and applicability than to the originality of the material. In presentations, offer relevant historical background and evaluate the relation of your work to previous efforts. Make sure your audiovisual materials are easy to follow. People are turned off by illegible and sloppy artwork. You should be able to field questions from the audience, so anticipate their most likely reactions or questions.

Writing technical articles is another useful form of advertising. These papers build credibility in your specialty and add weight to your résumé. The articles you submit need be neither brilliant nor original. Leave that to the Ph.D. students. Aim at *useful* technical articles that give you exposure to potential customers. Many specialty journals and industry monthlies solicit survey and descriptive articles aimed at practicing engineers and technical specialists. These unrefereed publications are an excellent way for beginning consultants to get their names in print.

Participating on society technical committees is another means of indirect advertising. The American Society of Mechanical Engineers (ASME), for example, has technical committees in hundreds of specialty areas. Their purpose is to establish rules, codes, and standards. Many consultants develop a "presence" at these committees because it exposes them to interesting opportunities.

Companies experiencing problems with interpretation of the ASME codes often ask a committee for advice on a ruling. Technical committees thus provide the perfect vehicle to meet clients who have pressing needs. By virtue of your membership on the committee, you are already established as an expert. To find clients with this method, prepare for the long haul. It often takes two or three years for these leads to pay off.

Writing books, monographs, and computer software is also useful in establishing a reputation. However, be warned that writing can become an all-consuming task in its own right. Writing books and software requires a major investment of time and energy. Although a few individuals achieve best-seller status, the vast majority of authors find royalty income much smaller than expected. I don't discourage you from writing books or software, as long as you are clear about the objectives you hope to accomplish in doing so.

The last form of indirect advertising is having a track record of successful projects and a set of good references. Toward this end, ask for a written endorsement after every successful consulting project.

 Tip

Request your endorsement immediately after you finish a project. If you ask a year or two later, your chances of getting the client to exert himself on your behalf are substantially lower. Be explicit about what you need: a signed letter, on his letterhead, stating that he was happy with your consulting service, that you handled the job capably, and that you gave him good value for his money.

FIND OUT WHO YOUR CUSTOMERS ARE

Cold calling is a standard sales technique in many industries. Insurance salespeople use it, for example, because nearly everyone needs insurance of some sort. They simply call every number in the phone book and give a short "pitch." Experience shows that they can expect a small, but not negligible, percentage of their audience to show interest.

For technical consultants, cold calling is an ineffective marketing tool. Your customers are more like needles in a haystack, and your sales efforts will be wasted unless you use a different strategy. *Prospecting* involves research about companies, individuals, new technical issues, sources, suppliers, regulatory agencies, and anything else that relates to your customers' technical needs with an eye toward finding those needles in the haystack.

This stage of marketing depends heavily on your ability to do good research. The consultant who has the best information — and who uses it — has a significantly higher chance of getting the contract. Prospecting for new clients involves a lot of legwork. You need to know: What companies are doing work in your technical areas? What specific projects or products are they working on? How great are their needs? How have they satisfied the need for specialized technical help in the past? Do they have a timetable to finish the project or product? Is the client under financial pressure to solve the problem? Who is in charge of your technical area? What is the company's history with respect to hiring consultants?

Don't expect to find this information conveniently assembled for you. The first step in prospecting is to garner the facts from such indirect sources as

- Stock reports
- Advertising brochures
- Government agency reports (for example, if you are an environmental consultant, look up recent EPA citation reports to find out if a prospective client was recently fined for waste discharge violations)
- Gossip from suppliers who deal with the prospective clients
- Reports from regulatory agencies that have technical jurisdiction over the product or project information from your client's customers
- Inside information (individuals who work for the client, at any level, can help you locate key people and confirm news items)
- Newspaper and journal articles (but be aware that by the time the media publicize a situation, the time for marketing consulting services may already have passed)

- Rumors from the clients' competitors (be careful how you proceed with this avenue, because if a client suspects you are dealing with a direct competitor, the door will be slammed in your face)
- Employment advertisements and jobshop requisitions for contract help. Clients often try to hire a jobshopper at a significantly lower rate than a consultant would charge. Yet, sometimes, the project falls beyond the capabilities of the particular jobshopper. After a month or two of fumbled attempts, the urgency of the project deadline requires plan B — hiring a consultant. I always keep track of advertised jobshop positions for this reason.
- Hearsay and verbal reports about start-up companies in your technical area

DEFINE YOUR MARKET TARGETS

The second step in prospecting is to define your market targets. The key to this process is *segmenting the market* by

- Industry
- Ultimate customer: retail, wholesale, distributor, agencies, individuals, companies, nonprofits, educational institutions, etc.
- Demographics
- Geographics
- Buying history
- Personal characteristics
- Identifiable situations (for example: crisis, corporation building a new facility, moving to new town, new job, college graduation, company obtaining large contract)

If your target customers are individuals (as opposed to organizations), be aware that there are two different "economies" in the United States. The first, and most visible, is the corporate economy; the second is the consumer economy. In marketing your consulting services, remember that companies have a lot more money to spend than most individuals. In terms of money,

1 Personal Dollar = 5 Corporate Dollars.

Corporations think little of spending $10,000 on office furniture for a midlevel manager. They will spend $20,000 on a three-day quality training seminar for a group of six employees. Few individuals are able to spend their personal resources at this extravagant rate.

Warning

If your consulting practice targets individuals as clients, do sufficient market research to establish the prices that your market will bear. You can easily get a company to pay you $150 per hour, but many individuals balk at even $40 per hour. To improve this situation, offer "package deals" (priced on a per-job basis) that make your service more affordable to a broader spectrum of customers.

HOW TO QUALIFY PROSPECTS

In *Marketing Your Services* (see the Suggested Reading List), Anthony Putman makes the point that you can't be successful taking whatever work comes along. He calls this a "Zorba the Professional" attitude. It leads to professional dead-ending because you can't create enough *focus* this way. Marketing requires many decisions. Some of these are tough decisions, because focusing energy on your prime targets means letting go of peripheral objectives. You simply won't have the time, energy, and resources to simultaneously pursue all options.

Qualifying prospects helps in this regard by letting you zero in on the target. You get closer and closer to the very people who will ultimately become your customers. The qualification process creates an orderly means to find new prospects and separate those who are only casually interested from those who are ready, willing, and able to use your services.

Criteria for Qualifying Prospects

1. Does the company have a pressing need for your service?
2. Does the prospect have decision-making power with respect to hiring consultants?
3. Does the company have the means to pay your charge rate?
4. Does the company have a history of hiring independent consultants?
5. Does the company have a good reputation in the industry?
6. Is the prospect willing to keep you briefed on developments in her company?
7. Do you have something to give the prospect in return? (There must be some reciprocity to maintain the relationship.)

8. Has the company recently received large contracts or committed to major internal development projects?

9. Is the company's need in direct support of its major service or product line?

10. Will many competitors be seeking this opportunity also? Is the prospect locked into a consulting arrangement with one of your competitors?

11. Is this a large contract with follow-on possibilities?

12. Do you have a unique selling advantage in this situation (for example, special equipment, customized computer programs, security clearances, or special professional licenses)?

13. Do you know the company well, or are you an outsider?

14. Will the company make a good reference? Can it open the door to *other* clients you want to meet?

15. Is it geographically convenient?

16. Do you already have credibility with the company?

17. Is the company's response time compatible with your timetable? (If you must wait eighteen months for a prospective client at a government agency to cut you a contract, you could go out of business before your ship comes in.)

THREE PRINCIPLES FOR REACHING PROSPECTS

Once you know your targets, you need a way to catch their attention. The best way to do this is to offer *something that the client wants*. From a marketing viewpoint, reading about the features of your service is not enough to catch the client's attention. To catch someone's attention, you must consider *his* needs and desires, and phrase your pitch accordingly. The client must be able to recognize, "Hey, this guy is talking about *me* and *my* exact problem!"

For example, "John Smith offers expert cryogenic design services" could be rephrased to "Are you struggling to find the right data for your cryogenic design?" And "Bob Jones has three degrees and twenty years' experience in vibration testing" could be turned into an attention grabber with "Are vibration tests busting your budget?"

1. Give them something they want and will keep. For example, a technical reprint, a data sheet, a sample, or a freebie. This Cracker Jack philosophy works with technical consulting as well as candy-coated popcorn, as long as you offer an appropriate toy with your package.

Newsletters are a wonderful prospecting tool in this regard. They are easy to produce with inexpensive desktop publishing software. The minimum practical size is one double-sided sheet, but four-page newsletters (on a single folded eleven-by-seventeen-inch sheet) are more attractive. The topics should cover the range of your specialty offerings. Include industry news items, articles of interest to people in your technical specialty, data sheets, and reviews of books and computer programs in your specialty. About 80 percent of the content should be externally focused. Allocate the remaining 20 percent to self-promotion: short articles describing your own research, contracts, and successes. Timeliness and reader interest are more important than originality. Make sure your name, address, and phone number are included so that readers can contact you.

Give the newsletter to anyone who sends for it. It's a lot of work to do this every month, so aim for quarterly or semiannual publication. After sending three issues to an individual, enclose a coupon requesting that the reader confirm continuing interest in receiving the newsletter.

2. Include a means to ask for more. With the free item, include a coupon that encourages prospects to ask for more data: another reprint, a demo disk, or a free estimate. Be sure your company's name, address, and phone number are included so they know how to reach you. Having them fill out the coupon is a low-pressure way to get their address and phone number. Additionally, it further qualifies the respondents as sales prospects.

3. Always structure a next step. Lead prospects so gradually that they are not aware that they are being led. Don't give them all the goodies at once, but one at a time. This way, you can further qualify each prospect and create an excuse to keep up the contact.

Structuring the next step takes *foresight* and *creativity*. Consider how Roger, a consultant whose specialty is cryogenic mechanical design, does it.

When Roger gives his annual ASME seminar on this subject, he casually mentions, "Anyone who would like a printed copy of today's presentation, please give me your business card with 'Mail ASME paper' written on the back."

Roger then mails each prospect a copy of the presentation and a coupon to receive a free data sheet that would be of great interest to anyone doing cryogenic mechanical design. If and when the prospect returns this coupon, he sends it, along with yet another coupon offering additional data. By the time the prospect sends in the second coupon, Roger is ready to call and propose a sales meeting where he can *personally* deliver the second set of free data.

Roger comments, "The first data I offer are pretty standard stuff. The information can be found in reference books at any good university library. The second set of data is more specific and would be of interest only to clients who can use the kind of consulting I offer. When I make my sales call, I offer the really marketable stuff—my own data. I explain to them why it is unique and how they can benefit from my consulting."

Once you receive responses from your prospects, sift through them to find the ones that meet *your* needs and criteria. First, filter out the responses that indicate casual interest, people who have no real need for your consulting. At this stage, also filter out responses from competitors who are curious about your marketing strategies. (Never mail your marketing materials to competitors!) You can safely discard individuals who have inquired to learn more about you as a potential employer. Likewise, accountants and equipment vendors will try to qualify *you,* the same way that you are targeting and qualifying *your* customers. All these prospects can quickly be dismissed because they cannot possibly be customers of yours.

In the second sift, filter out the prospects who are interested and who are in your target market segment, but who do not have a need within the next six months. At this stage, also filter out prospects whose companies do not hire consultants as a matter of policy. The prospects in this second filter are long shots. Marketing them will require significant work and time.

What remains after your second sift are your *qualified candidates,* your best bets. This group of prospects has immediate and identifiable needs in your target area and is inclined to do business with an independent consultant. After doing your research, contact them by phone and ask to meet with them. Set up a sales meeting to further discuss their needs.

At each point in the qualification process, develop the criteria and strategies to bring prospects along to the next stage. Typically, twenty prospects produce one qualified candidate, so the more efficient you are at sifting, the better your results.

TRACKING YOUR PROSPECTS

To keep track of your prospects, use a database or a contact management program on your personal computer. The latest contact management software allows you to search through your data efficiently and make convenient summary reports of your findings. With it, you can monitor your marketing efforts and results. The price for these programs has been dropping steadily in the past few years; you should be able to find one for under one hundred dollars.

Contact management software allows you not only to store names, phone numbers, and addresses, but also to maintain a detailed log of phone conversations and action items. Most contact managers have blank (user definable) fields that you can adapt to the particular needs of your consulting practice.

Perhaps, for example, you have six primary industries that use your service, and you want to track which industries are the most fertile ground for developing clients. You could ask the software to sort based on industry and provide totals to get a quick answer. Developing this kind of information manually is time-consuming and error prone.

You might want to use another field to qualify the prospect, ranging from "casual interest" to "red-hot." Other items you may want to include as database fields are:

- Which materials you have sent to each prospect
- When and where first contact was made
- Who referred them to you
- Dates for follow-up
- Their initial reaction or comments

One word of caution: Use software that employs industry-standard file formats (such as dBase's .dbf or Access's .mdb). You then can then switch or upgrade programs, if need be, without having to reenter all your information. Reentering information for a large database can be extremely time-consuming. The cost of the software itself is negligible compared with the labor cost of data entry.

Most database and contact management programs allow you to compose form letters and select individuals on the list to receive them. Once you set this up, the program automatically inserts the client name and address for each letter and prints mailing labels—a great time saver.

WHY YOU MUST MAKE PERSONAL CONTACTS

Marketing new clients is generally accomplished through *personal contacts*, or "connections." Clients are wary of hiring a consultant who is not a known quantity or who does not come referred through a trustworthy source. Advanced degrees, titles, and professional licenses are no substitute for the consultant's personal contact with the client. The consultant's foremost marketing effort is, therefore, to *become known as a credible and cost-effective problem solver* to as many clients or clients-to-be as possible.

You have probably already developed some of your contacts. In your career as a direct employee, you have established yourself as credible and competent among some of your co-workers and your company's suppliers and customers. These people form the initial core of acquaintances with whom you should keep in touch. Over a period of years, these acquaintances will move to different companies, where they may be in a position to use or recommend your services.

Contacts are not the same as prospects! Prospects are your potential clients. In contrast, contacts can come from any social segment. You never know when or where you will meet a useful contact. I have met some seated beside me on an airplane flight. Relatives, club members, church members, and fellow consultants are also good candidates.

Prospects are more specialized because they fall in the target market segment at which you are aiming. Although it's possible to meet prospects anywhere, it is far more likely you'll meet them where they congregate in numbers. This is why so many consultants go to professional meetings and offer seminars that appeal to their target market segment.

You can make interesting contacts at the consultants' networking groups established in larger cities across the country. Since the people there will be *sellers* of services similar to yours, it's unlikely you will find *buyers* of your services (prospects) there. *Prospecting is finding buyers, not other sellers.* Accordingly, adjust your expectations when you are networking. Don't confuse it with prospecting. Contacts may not be future clients, but they are the ones who help to open the doors to future clients.

Learn to make new contacts at professional meetings, at interface meetings with your present employer's suppliers and customers, and even at social gatherings. The aim of these contacts is to

1. introduce yourself as a technical consultant;
2. establish yourself as knowledgeable in your field (in a subtle way, of course);
3. determine the suitability of the other person to advance your own causes, in terms of company position, social circles, business contacts, etc.

In particular, look for chances to meet people employed by (or connected to) your target companies. Of course, you should expect some resistance to such efforts to make contacts. Five-star contacts are like extremely attractive movie stars. They *know* that their attention is widely sought. They *expect* vendors to be solicitous. They will be hard to approach unless you have something valuable—in their eyes—to offer. Therefore, don't be upset if your efforts at making the "big" contacts are thwarted. It says more about the other person's need for privacy than about your marketing ability. By not becoming upset at the dozen noes you may receive at their hands, you will be able to arrive at a yes that pays off.

Making good contacts takes great listening skill and intelligence so as to leave a good impression without making the person feel that you view him solely as a means to your business ends. Establishing rapport is much more important than trying to impress the contact about your technical prowess.

KEEP CONTACTS "PROFESSIONAL"

Meet with your contacts occasionally. Stop by to show them your new brochure or to have lunch together. Do not try to sell anything in these meetings, and keep the tone light. Be sure to mention your latest technical interests with enthusiasm. It may help to send your contacts technical articles that are of interest to them. Or, call to make them aware of a professional meeting that may have escaped their attention. The idea is: Stay in touch on a *technical* level.

As a consultant, the image best maintained is different from that of a corporate manager. In large companies, managers commonly gain rapport with one another through golf, tennis, and weekend socializing. This is fine for them, because these activities cement an already established bond: the parties work for a common employer. But for the consultant, no such common bond exists. It is far more likely that you will get a contract from a client because he thinks you can solve his technical problem than because you are drinking buddies on the nineteenth hole. That is, let the contact *think* of you as a consultant, and not as a "76 at Torrey Pines."

Do you *need* to treat clients to lunch and send them Christmas gifts to get ahead in consulting? The answer is an unequivocal no. I don't have firm rules about treating and giving gifts to clients. I follow my intuition. For example, a number of years ago, I gave one client a nice bottle of cognac at Christmas, even though his company sent a form letter to all vendors and consultants, which insisted that I *not* give its employees Christmas gifts. That client had gone out of his way to help me get a contract with his company, and I simply appreciated it.

At business dinners, only occasionally do I feel the need to pick up my client's tab. If the focus of your client relations is going out to fancy restaurants with you picking up the tab, then watch out! More often than not, the real performers among your client's staff are busy people also. If a proposal meeting runs into lunch hour, they will have their secretary bring in sandwiches or suggest you accompany them to the company cafeteria. I never refuse such suggestions. Each additional minute you spend at the client's facility will expose you to more of the staff and the way the company operates, all of which provides valuable information.

CONTACTS ARE THE "SEEDS" OF YOUR CROP

Professional contacts are the "seeds" you plant to reap your "crop" of consulting work. Just as the farmer orders and stores his seeds with due thoughtfulness, you will need to acquire and maintain your contacts.

The analogy goes further, though:

- Planting seeds is essential to the farmer's success. Contacts are essential to the consultant's success.
- The farmer recognizes whether the soil is appropriate for the type of seeds he is planting. The consultant knows which technical markets are fertile for the type of contacts he has.
- Not all seeds germinate. The farmer focuses on the garden, not on the individual seeds. Likewise, not all your contacts will sprout into consulting work.
- Early in the season, the farmer eliminates weak plants and weeds. He then concentrates on the strongest and best-yielding plants. Likewise, the consultant cultivates his contacts for higher yield. To do this, set specific goals for your networking. ("I want to meet six people next month, three of whom I hope to eventually bring into my network.")
- The farmer uses new seeds and revitalizes the soil annually. Every year, the consultant makes new contacts and revitalizes himself technically.

HOW CONTACTS WORK

In previous sections, I've explained that contacts are an important part of marketing. That's great in theory, but how do contacts work in practice? To illustrate the process, let me tell you how I got the multifaceted contract mentioned in Chapter 2. I've already related how my unique qualifications in that situation allowed me to beat the competition. What I haven't described is *how* I went about it.

I was having a casual lunch with a business friend. During the course of our conversation, he mentioned that the XYZ Company was trying to find a consultant in my specialty. In fact, the company's purchasing department was circulating, on a very limited basis, an RFP (Request for Proposal) for specialized technical help in gas dynamics and heat transfer. I asked more questions, but that was all my friend knew. Moreover, he had no contacts at that company.

When I got home, I called Bob, the one person I knew at the XYZ Company. Did he know the group that had issued an RFP for help in my technical specialty? Well, he didn't, but he would ask. Two days later he called back, saying that he had asked half a dozen of his friends, but none of them had heard of such a project. Maybe, he suggested, I had misunderstood my friend, and the need was at some other company. I was crushed. Bob was my only contact—or so I thought.

Another day went by, and while I was taking a shower, it occurred to me that I had another friend, Frank, who was not working at XYZ but who was involved in the same technical area. I called Frank and asked him if he knew anyone at XYZ who might be developing a new project using my skills. He responded immediately, "I don't know what new projects they are starting, but I do know Lew B. really well. He's their director of research and development. If anyone would know, it would be him. Just tell him I told you to call."

I called Lew B. and introduced myself as a friend of Frank's. I then went on quickly to say that I was an independent technical consultant in heat transfer and gas dynamics, and that I had worked at the ABC and DEF companies in the local area, where I had met Frank (our mutual acquaintance). Lew asked me if I knew Jerry S. at ABC, and I said yes—Jerry was in charge of a certain development project. I explained that I had helped Jerry in some heat transfer calculations on that project. At this point, Lew knew that I was not just anybody calling him on the phone. He had sufficient information to feel comfortable opening up to me. I told Lew that I had indirectly heard XYZ was trying to obtain help in my exact specialty. Could he verify this and direct me to the right person? "Well," he said, "I think our new laser development project is what you're describing. Why don't you call Henry F., the project manager, and see what he has to say?"

Finally, I called Henry F. Again, I introduced myself by mentioning that I had just talked to Lew B., and told him my consulting specialty, that I had worked at ABC and DEF in that technical area, and that my graduate school background at M.I.T. was also in that area.

I said all this in a relaxed manner, but I was careful not to go into detail. I was casting out three "lures" he could bite on, but I didn't want to overdo it. When you are talking to a person without benefit of a formal introduction, you can get in about four sentences to introduce yourself. After that, you are trespassing beyond the bounds of common courtesy and patience. Therefore, get as many lures out in those four sentences as you can.

Henry bit on the M.I.T. lure. "Gee, I went to M.I.T. myself. What department were you in?" Our conversation took off from there and lasted twenty minutes. There was indeed a need for help in my specialty, and the purchasing department had just made up an RFP. Henry would have them send me a copy. Further, I told him I wanted to set up a face-to-face meeting to show him a portfolio of similar projects I had worked on. Henry agreed, and the rest (as the saying goes) was history.

USING CONTACTS

- Understand that some people don't even know what's happening in their own company. And others may be reluctant to share their information with you.
- Always ask, "Who else might know? Would you introduce me to him or her?"
- The phone is the best medium for quickly zeroing in on a target. Letter writing and E-mail take too much time.
- Be brief. Don't inflict lengthy inquiries on your contacts unless they "owe" you.
- Throw out as many "lures" as you can.
- If you don't succeed at first, try another approach. Don't give up! Be creative. For example, maybe you can reach the target person through a supplier, a tennis buddy, or another consultant.
- Return the favor when your contacts require something of you. If there is no mutuality, the contact will eventually fail.
- Contacts are based on goodwill. Don't put your contacts on the spot or pressure them.

HOW MANY CUSTOMERS DO YOU NEED?

A consulting business can be started with only a few steady clients. Once your practice is established, make it a priority to increase the number of clients on your roster. This allows you to move around and prevents a locked-in feeling. It is easier to plan your future knowing that your fate is not tied directly to the fortunes of just one or two clients. Spreading out your client base also reduces the pressure to prolong contracts or milk them. Finally, moving around forces you to stay efficient and dynamic.

On the other hand, if you have *too many* clients, there is danger of spreading yourself too thin on the technical issues relevant to each one. Also, too many clients may call at the same time. Or you may have too many of your proposals accepted within a short time span. At this point, you may be ready to hire employees or subcontractors of your own to help with the work load.

The happy medium depends to a certain extent on the nature of your consulting specialty. If you are performing a service that can be done in a few hours, you will need dozens of customers. On the other hand, if your typical contract involves months of effort, you can make do with a handful. In the

Tip

In writing proposals, allow the possibility that some of the work may be done by assistants or others on your staff.

beginning, err on the side of having too much business rather than not enough. This not only generates more cash flow, it also provides the base of experience you'll need to optimize your consulting practice. You will find it easier to drop clients who are a poor match to your goals and add clients who are a good match.

When you are busy and working forty billable hours per week, what do you do when another client approaches you and asks for your time? If you cannot negotiate the starting date of the new project to begin after your present commitment ends, you must make a trade-off: Is the new opportunity worth working overtime to satisfy two clients at once? For me, it usually is. I occasionally work sixty or seventy hours per week in such situations. In most cases, the overlap period is only a few weeks. This temporary disruption to my evening and weekend schedule is bearable. Once I start working with the new client, I seek ways to postpone or delegate certain tasks, thereby recapturing some of my overtime.

I back down from double booking when it is clear that the situation will last for months. Long-term overtime is a sure prescription for burnout. In this case, I try to sell the client the services of a subcontractor who works for me.

LEARN WHICH COMPANIES YOU CAN DO BUSINESS WITH

 ## Worth Its Weight in Gold

Here is a marketing secret that can be worth a hundred times the price of this book: *It will be extremely difficult to obtain consulting work from certain companies, regardless of your talent or their degree of desperation.* This is true for any number of reasons:

- They have a policy of not hiring consultants except in dire emergencies.
- The prima donna attitude: They know all there is to know, and outsiders couldn't possibly know anything more.

- They see you as a potential competitor.
- They are concerned that you may take proprietary secrets. (At times, even a proprietary interest agreement does not sufficiently protect valuable secrets.)
- They don't like you, because you don't have a Ph.D. (or because you *do* have a Ph.D.!).
- They may have an exclusive or long-standing arrangement with another consultant to provide the very services you are proposing.
- They don't care to do business with a person of your race, gender, color, or religion. While certain laws protect the public from discrimination on this account, they are not really effective in the consulting business.

The value of this secret is not knowing that such clients exist, but knowing that as soon as you have determined that a client falls in this category, you needn't waste any more energy on them. Save your precious marketing time to pursue other clients. Although most disinclined clients will brusquely refuse to see you, a few will take you on an "amusement ride." Without any intention of doing business with you, they "dangle the carrot" in front of you and solicit your free advice. They may even ask for a quotation on a project. But they never come through with a paying contract!

What's the best way to determine whether you can do business with a given company? First, find out (from contacts and friends) if the company has used other consultants in the past. Second, call up the purchasing department or contract labor administrator and ask if there is a list of approved consultants or contract labor vendors. Find out what it takes to be put on this list. And finally, call the manager of the department most likely to use your services, and give him your brochure (in person, if at all possible).

By the time you have taken these steps, you will know whether you are marketing a company you can't do business with. If you are, forget it. Concentrate on customers with whom you have greater chances of success. Remember, there are thousands of clients out there; you only need to connect with a few to achieve your consulting goals!

OTHER CONSULTANTS MAY NEED YOUR SPECIFIC TALENT

Folklore has it that consultants in the same technical area are competitors. It doesn't have to be that way. Often it's possible to establish a mutually beneficial relationship with another consultant or consulting company. I have worked as a consultant to other consulting companies. If it's OK with them, it's OK with me. However, there are three issues I establish before I make an arrangement:

1. I get my usual billing rate.
2. The consulting company agrees not to represent me as a direct employee to its client.
3. I agree not to market the company's client while I am working on its behalf (and often for a period of time thereafter).

These arrangements have led to substantial new opportunities that I could not have arranged by myself. As an example, about five years ago, I was approached by friends at a medium-size consulting company. They needed help with an urgent problem in transient two-phase flow. Their client was a large power company that had given them an umbrella contract for engineering support. It would have been impossible for me to obtain this work by approaching the power company directly; it dealt with megawatts (and megabucks!). It didn't want to be bothered with small contracts and the administrative hassle of a hundred individual consultants. The consulting company profited by being able to satisfy its customer's important problem, and I got to work on an interesting technical problem.

HOW TO GET EXTRA MILEAGE FROM A CONTRACT

In marketing, one contact leads to another. Be sure to realize this potential as you perform your work. I have had the frequent experience of being hired by a client to do a particular job, and meeting people at that client's facility who had more projects on which I was later asked to participate. How do you manage to get yourself in such situations?

- Do excellent work.
- Seize the opportunity to present your work at your client's internal review and interface meetings. Be eager to participate. Put in the extra effort to use well-prepared and attractive presentation materials, even if the manager in charge says it's a very casual meeting.
- At the completion of a project, prepare a memo with neat graphics and a clear explanation of your work. This becomes, in a manner of speaking, your "autograph."
- Be open at all times to meeting other people on the client's staff who are involved in other projects. Always carry your business cards in your wallet for this purpose.

MARKETING THE PUBLIC SECTOR

Many federal, state, and local governments have requirements for technical consulting. The U.S. Census Bureau cites approximately 80,000 federal, state, and local government agencies that are potential buyers of consulting services. The combined governmental dollar volume of business is huge (about $800 billion per year). Of course, the largest part of this figure is for materials and equipment, such as major roadways, new planes for the Defense Department, and giant computers for the Census Bureau. Hidden in this market, though, are still *billions* of dollars for services that consultants perform: preparing feasibility studies, collecting data, making evaluations and recommendations, training, writing reports, maintaining technical equipment, and performing efficiency studies.

The trick to getting government contracts is to understand the procurement processes of the individual agencies that offer them. By law, the government is required to give part of this work to small businesses.

To investigate the federal sector, get a copy of the Commerce Business Daily, published five days a week by the Commerce Department. This newsletter lists the needs of many federal agencies and gives information on obtaining bidders' packages with further details. For a sample copy of the *Commerce Business Daily* and ordering information, write the GSA (General Services Administration) at 18th and F Streets NW, Washington, DC 20405. To view a searchable on-line version, see the *Government Contractor Resource Center* Web site at http://www.govcon.com.

The Small Business Administration (SBA) can also help point you in the right direction. It has offices in many large cities; its headquarters is at 1441 L Street NW, Washington, DC 20416. The SBA can also be reached on the Internet at: http://www. sbaonline.sba.gov.

As you can imagine, doing business with government agencies has its drawbacks. Bureaucracy, time delays, and competitive bidding are significant challenges. Expect to fill out many forms certifying compliance with federal laws on equal opportunity, hiring practices, supplier policies, and so on. For consultants with an inside track, though, providing services to the government opens many new markets.[1] Further, if the contract value is under $10,000, less paperwork is required and a simple purchase order can be issued in place of a formal contract.

---

[1] For an inside view of the government contracting business, I refer you to Herman R. Holtz's book, *Government Contracts: Proposalmanship and Winning Strategies,* Plenum Publishing, New York, 1979.

THE INTERNET: YOUR KEY TO THE WORLD

As this third edition is being written, the Internet is experiencing explosive growth as a new communication medium. I have used the Internet for the past few years and find it incredibly useful for E-mail and doing research for marketing, books, and articles. The potential for further developments is extremely rosy. In fact, we are beginning to experience *digital convergence,* a combining of technologies that were previously separate. In the near future, the Internet — and its successor technologies — will merge computer, phone, fax, television, CD-ROM, VCR, and many other separate devices into a global communication system with real-time transmission of voice, data, and video channels.

On the one hand, I am reluctant to write about the Internet here because it is a foregone conclusion that developments are occurring so fast that whatever I write will be obsolete within two years. On the other hand, I would be derelict in my responsibility to you to ignore these significant and exciting innovations. So please bear with me.

The Internet is invaluable to consultants for

- *Advertising:* Create your *own* Web page as an on-line brochure that transcends traditional paper formats by adding sound, animation, hyperlinked texts, and even interactive programming. The cost of "printing" is zero, and your potential exposure increases by a factor of 1,000.

- *Marketing:* Some companies have Web home pages with job listings for direct employees. This gives you an excellent opportunity to understand their detailed technical needs. A fraction of these companies will be receptive to your counter-offer of technical support on a consulting basis.

- *Contacts:* Most companies and many individuals have their own Web pages that allow you to contact them quickly and easily. Many companies also have personnel directories that permit you to zero in on the individual you want to contact.

- *E-mail:* E-mail is faster and cheaper than the U.S. Postal Service ("snail mail"). I find that clients and individuals who don't know you tend to be more accessible through E-mail, as long as you are succinct.

- *Reference:* Throw away that ten-year-old encyclopedia and those dusty reference books! The Internet is a great research tool right now and is destined to become much better in the future. With enhanced search facilities, you can find the precise material you are looking for.

- *Professional groups:* With Internet chat and news groups, you can join a "virtual" professional group or create your own. As with any face-to-face group, be cautious and prudent before you reveal personal information or commit your time, money, or support.

MEET THE PROS UP CLOSE AND PERSONAL

Martin Madigan is the president of Quality Engineering Services in Macedon, New York. His consulting is in the field of statistical process control, where he has spent the last twenty years applying statistics to manufacturing processes for large industrial clients. His  reputation as a seminar leader is worldwide; he has consulted and lectured in Mexico, Brazil, Germany, the Netherlands, Singapore, Hong Kong, Korea, and Canada, as well as throughout the United States.

Marty says that if your consulting has been mostly stateside, you need to be aware that a different set of rules apply to international business. The first new rule is: Watch out for the water! Nothing is worse than arriving in a new country and developing a case of dysentery just before you are to deliver an important seminar. Second, stay sober! The laws for drunkenness are different in other countries. You can place yourself in great risk by drinking too much. Third, take the time to learn about the country's customs. In Singapore, for example, when you ask someone, "Do you understand?" an up-and-down nod means he or she *doesn't* understand. Or, in Brazil, when you indicate "A-OK" by holding your first finger to your thumb in a loop, Brazilians take that gesture to mean, "f— — you!"

When Marty goes abroad, he carries very little cash. He says, "The world has become one large credit card community. All I need are my VISA and MasterCard." At Marty's insistence, all his contracts are written to provide for payment in U.S. dollars. His clients usually agree with little fuss.

Marty says that international business also follows different standards for travel. He insists on traveling first-class to most far-off destinations and advises other consultants to do the same. Traveling economy class for long distances will break your back and frustrate you with inconveniences. Marty charges for travel time, based on actual flight time. The first eight hours of flight time are billed at straight time. He charges time and a half for anything over the first eight hours. Thus, a flight to Hong Kong, which takes twenty-four hours, is billed out at $8 + 1.5(16) = 32$ hours. It doesn't take too much long-distance flying to realize that it is extremely exhausting. His clients appreciate this also, for Marty says they almost always accept his billing terms.

Moreover, through long experience, Marty realized that being in a foreign country for a week or two at a time precludes doing any other work during off-hours. As a result, he charges eight hours per weekend day as well, even when he is not consulting or teaching.

GLOBALIZATION

You don't have to use the Internet very much to realize that it is promoting the creation of a *global* community. Companies all around the world are talking to one another more freely, exchanging technical information, and offering products and services internationally. You can, too!

Think about it: if you allow the possibility of global business, your potential market increases by a factor of ten to fifty. I know many consultants who find it easier to get lucrative consulting contracts outside the United States than inside. With a little more research and networking, they escape many market territories that are saturated in the States but underserviced in foreign countries.

These international opportunities are well worth exploring, as long as you remember that you have extra paperwork and overhead costs when you do business across the border. In most cases, add a surcharge to your regular rate (described in Chapter 9) to recover unbillable expenses such as foreign taxes and licenses, passports, inoculations, money conversions, and other incidentals of international travel. Also, be aware that cultural differences, banking standards, and legal restrictions vary from country to country. Allow some space on the learning curve to accommodate these new factors.

GET YOUR FOOT IN THE DOOR

The goal of all marketing work is to meet your future clients face-to-face. In most cases, you will be the one who makes the approach. When you call a qualified prospect for the first time, do not talk about yourself until you're asked to. Instead, discuss the *direct benefits* to the client that will result from your interaction.

At other times, your future clients will be calling you. It is *very* important to recognize the potential that lies dormant within every "casual" inquiry.

When an acquaintance asks for technical information, do not hesitate to look up a reference or do a quick back-of-the-envelope calculation. These inquiries are your best marketing leads and must be handled with tact and competence. If you can't give a simple answer on the spot, tell this prospective client how one would go about solving the problem in general. That is,

Tip

Get your bait out there in the *first thirty seconds* of your call. For example, "Hello, my name is John Smith, and I'm a consultant in mechanical vibrations. I understand from a mutual friend, Lee Jones, that your group is in need of someone to write the vibration test specification for your Delta project. Would you allow me a few minutes to explain how I can *save you a lot of money* using the special techniques I've developed over the past ten years?"

outline the major steps of the solution. You must make the other person feel that you are interested in his problem and that you are competent to solve it. Don't worry about giving out free merchandise. Think big.

In a first inquiry, the client may ask if you have the time and inclination to do the work. Before you answer, try to establish the *time frame* of providing the service. Does the client need your assistance right away? Or is the client asking for help to be rendered next year? Determine if there are any restrictions on *where* the work must be done. Further methods of qualifying inquiries are given in Chapter 7. For the moment, here is my advice for dealing with hot leads:

1. It is best not to be available "right away" unless you have done business with the client previously and know the business arrangement. For new clients, it usually takes weeks to get rates, purchase orders, and contractual matters lined up before you actually start work.

2. If your client is a small company (sales under $20 million), wait for a *written* purchase order before commencing work. It's sound business practice.

3. Establish whether you are talking to the person with the authority to approve your involvement. This is extremely important. I wish I had a dime for each time a subordinate told me, "Save next week for us," and was then vetoed by a supervisor who thought otherwise.

4. A client may just be "window shopping" for ideas on how to solve his problem, without any intention of hiring anybody. This is business. You might shop for a television in this manner yourself. Be as courteous and helpful as you can—the first few times. After that, if it seems that he is trifling with you, drop him in the "can't do business with him" bucket.

5. Ask questions that enable you to assess the nature of the problem and that may uncover useful background information.

6. Resist the impulse to estimate the scope of the work until you have all the necessary information at hand.

7. If the client asks you to travel to discuss a potential proposal, evaluate the return on this investment. As a matter of policy, I will not travel a long distance without compensation, unless I know that my chances of winning a substantial contract are very high.

HOW TO CONVERT AN INQUIRY INTO A SALE

When handling an inquiry for technical information in my specialty, I use my greater knowledge of the subject to suggest ways in which I can help. For example, a few years ago my friend Steve called me. The conversation went something like this:

"Hey, Harvey, you don't happen to have a handy formula to estimate the heat transfer to a missile traveling at hypersonic speeds, do you? We need just a little hint to point us in the right direction."

"Sure thing, Steve. I have a whole slew of hypersonic heat transfer formulas I can give you, complete with references and derivations. But first tell me a few of the problem parameters so I can give you the best formula for your application."

Before I finished my next sentence, Steve interrupted with what he thought I meant by problem parameters. "Great. Here's what we have..." (Steve went on to describe the project in general.)

"OK, Steve, can you tell me the shape of this missile, its trajectory parameters, its surface material, and..."

Steve then told me the information I needed. At this point, Steve was so interested in getting his handy formula, he was unaware that we had established *technical rapport.*

"Steve, what you want is Equation 27 in Chapter 6 of the *Handbook of Heat Transfer* for stagnation-point heat transfer in dissociated air. This is exactly what you need because...By the way, who's going to do the calculations?"

Steve responded, "Oh, Len C. is going to do them. He got his B.S. last year from the University of Maine and has had two courses in heat transfer."

The person doing the calculations was clearly not a seasoned veteran in the field of hypersonic flow. This all made sense, for if anyone in Steve's company were sufficiently experienced in that field, why would Steve be calling me?

My mind raced to think of aspects of the problem that they might not have considered.

"Gee, Steve, do you know that the solution involves more than the formula? You'll need to compute various high-temperature air properties at each point in the trajectory, which is very tedious. I have a neat computer program that does this. If you use approximate properties [here is where I capitalized on my special knowledge] you will find that your answers can be in error as much as 60 percent. Why don't I help Len set up his own subroutine for air properties? It's only a few days' effort on my part."

Steve's tone of voice changed. He was hooked. "What is the earliest date you can do this?"

"I have to check my schedule, Steve, but tentatively I could save the nineteenth and twentieth of this month. [This was two weeks away.] Len and I could go over the basic equations for the properties and figure out how to integrate a simple subroutine for his specific application."

"Fine, I'll get Bob Jones, our purchasing agent, to send you a purchase order."

In this example, I already knew the client. I didn't have to sell myself, only the service. Under these circumstances, it was easy to "close" the sale over the phone. In many situations, though, you won't know the client, and you should maneuver strongly in the conversation to set up a sales meeting at the client's facility. Chapter 7 explains what to do once you've got your foot in the client's door.

CHAPTER 6

Your Blueprint for Marketing Success

This chapter is, in effect, a *Personal Strategic Marketing Session.* You could go to a high-priced business consultant for this, but here it is — read it in thirty minutes — included in the price of the book! What follows is a detailed blueprint for marketing success from ground zero.

Planning your marketing strategy is absolutely necessary in today's rapidly changing marketplace. You must become more strategically oriented as technical fields become more crowded and the pace of innovation accelerates. If you neglect this strategic orientation, you will wander from project to project — if you are lucky. If you are not lucky, you will quickly find yourself sitting on the career sidelines as more aggressive consultants grab the attractive projects right out of your hands.

Without a strategy, your marketing efforts will be scattered, because you will not know how to position yourself and capture the high ground in your niche. Without a strategy, you will have no sound reference point to make marketing decisions. You may possess a solid goal, but you will lack a defined path that explains how to get there.

The goal of all strategy is to set yourself apart from the rest of the pack in such a way as to gain an advantage. This advantage becomes what marketers call your unique selling proposition (USP). You should be able to state your strategy in three or four simple sentences. Keep it simple! If your strategy is complex and depends on too many factors, revise it until you find one that is simple.

You need only *one* good strategy to achieve success in the market. Of course, this one strategy may have substrategies that are required to carry out the details. Having too many primary strategies is like having no strategy at all — you will be at a loss in making tough decisions about where to place your marketing efforts.

Figure 2 shows the steps in the strategic marketing process I am about to describe. Just start at the top and work your down way to the pyramid's base. By the time you are through, you will have a good handle on setting your strategic goals and developing effective plans to achieve them.

FIGURE 2. The marketing pyramid.

FIRST VISION, THEN GOALS

The first step in developing a sound marketing plan is finding and acknowledging your passion. This means having a *vision,* a dream of where you want to be in five years. Your vision should fire your imagination and summon your emotional energies, making you ready to exclaim, "Yes! *That's* what I want to do professionally, *that's* where I want to go!" We already discussed vision in a general sense in Chapter 5. Here, vision refers specifically to strategic vision, that is, vision for the purpose of furthering your planning process.

Once you are in business for yourself, personal vision is the only thing you can rely on. Trust your vision! Other people's "sure bets" usually have a way of leading you away from where you want to go. Why? Because they don't factor *you* into the equation: They don't consider your strengths and weaknesses; they don't acknowledge the very things that excite you professionally.

Strategic vision helps you see *more options,* options that take advantage of your unique selling proposition. It helps you decide where, when, and how you should be focusing your efforts. Once you have developed your strategic vision, you will find it easier to answer the following questions:

- What new products should I be developing?
- How do I plan for expansion?
- What resources must I develop?
- How do I counter competitive threats?

- How do I gauge the attractiveness of new opportunities?
- What technological "swamps" should I avoid?

Vision precedes goal making. The pages of success magazines offer myriad examples of entrepreneurs who founded successful businesses based on *just one idea that they single-mindedly pursued.* That is, in our terms, a single strategic vision. So, before defining your goals (strategic mission), envision what you want your company to be like five years from now. Go ahead, be specific. State exactly what you want to be doing, and what situations you want to avoid. Dare to say "Who I am" and "Who I am not." Dare to say what your values are. Dare to say what you're trying to accomplish.

As an example, here is what the founder of Federal Express, Fred Smith, envisioned for his company:

Federal Express's Vision

- Provide totally reliable, competitively superior air-ground transportation of high-priority goods and documents.
- Maintain positive control of each package with real-time tracking and tracing systems that give a complete record of shipment and delivery.
- Strive to be helpful, courteous, and professional to each and every customer.

This vision statement is very effective because it

- is short and simple;
- shows the value that's being added;
- highlights the market advantages that distinguish the company from the pack;
- is specific in portraying a market niche.

Ideally, express your vision in the most vivid terms you can muster. Some thirty years ago, when Hichiro Honda expanded his motorcycle business to making cars, his strength was superior engine technology. His vision was to create engines with the lowest possible level of air pollution. To get this point across to his engineers, he instructed them to find a way to "take care of your own piss and shit." The engineers never forgot what old man Honda wanted from them!

Note that "I shall double my income in five years" is *not* an effective vision statement, because it doesn't portray you "doing your thing" or providing value to clients. For the same reason, "I shall leave my idiot employer and lead the life of a carefree consultant" is also lacking. It may be pleasant to entertain this thought for a moment, but it shows you escaping a bad situation, not moving positively toward a new one. Furthermore, avoid vision statements that are grandiose ("We are the best") or vacuous ("We want to be valuable to our customers.") Instead, focus on statements that translate into positive goals that can make a difference.

A vision statement does not state *how* you are going to attain the vision. It only provides a detailed image of what "doing your thing" looks like. The next step in the marketing pyramid takes care of the *how*. Strategic goals (your "mission") follow naturally from a clear vision statement.

Let's see how this worked at Federal Express. What were Fred Smith's strategic objectives (mission) in starting the company? Well, it turns out that he *already* had a brilliant strategy to implement his vision. As a graduate student at Yale, he had conceived a unique parcel delivery system based on a "hub-and-spokes" arrangement. Nothing traveled directly point to point; everything was routed to a central hub to be sorted and rerouted most efficiently. Smith submitted his idea for an economics paper and got a C for it.

Federal Express's Mission

- Provide competitively superior air-ground transportation of high-priority goods and documents by using hub-and-spokes concept.
- Maintain positive control of each package with real-time tracking and tracing systems by using the latest technologies.
- Focus on the overnight delivery market.

Ten years later, Smith started Federal Express. At first he tried to compete with Emory and Airborne by having three classes of delivery (overnight, two-day, three-day). It was a mistake, and losses in the first two years were $29 million. Smith then remembered Clausewitz's old rule about *focusing your attack* when going against stronger adversaries. Originally, Federal Express thought its mission included delivery of *all* packages. After two years, Federal Express changed its strategy and narrowed its market focus: "When it absolutely, positively has to be there overnight." By pushing the high-priority market segment, the company saw its profits skyrocket. Federal Express became a billion-dollar company soon after.

Your mission statement defines the specific strategies you intend to use in attaining your vision. At this level, it is important to keep the strategies broad and simple. You will translate them into detailed tactical imperatives later on in the marketing process. For example, here is how one computer consultant stated his mission:

My Consulting Mission

- Offer competitively superior client-server database systems using Visual Basic and Oracle as primary tools.
- Create a unique "audit program" to test and monitor the data integrity of the systems I am developing.
- Become an active member of society committee on data integrity and secure transactions to improve credentials and exposure.
- Focus on the banking market, where data integrity is paramount.

GATHER MARKET INTELLIGENCE

An army general would never plan a campaign against an adversary without first obtaining some military intelligence. He would want to know the enemy's deployment, strength, position, lines of supply, and combat readiness. He would want to know the terrain of the localities, the effectiveness of the troops' support systems, the political alliances that have been fostered, and most important, the *enemy's* strategic objectives. He would study its strengths and weaknesses in depth to discover chinks in its armor that would allow a possible advantage.

Similarly, as a consultant, you'll need to find out a lot about your market to make effective plans. Market intelligence has six dimensions for the technical consultant:

The Marketplace

- Who are the major players?
- What "drives" the market? Is it innovation, price, credibility, advertising, regulations, or performance?
- How is the market segmented?
- Is it a buyer's or seller's market?
- What are the "Rules of the Game"?

- Who sets performance standards?
- How are prices determined?
- How is acceptable quality defined?
- What is the basis to demonstrate superiority of your service compared with another vendor's?
- How are warranties made and honored?

Customers

- Who are your customers?
- What is their financial health and business outlook?
- What new projects are they developing?
- Why are your customers buying from you?
- What new products (services) do they need?
- What products (services) do they intend to *stop* buying in the near future?

The Competition

- Who are your competitors?
- How are they getting the job done?
- What are they doing better than you?
- What weaknesses do they display?
- How do they get their customers?
- How are they positioning themselves in the market?
- What resources and/or capabilities are they developing?
- What threats do they pose to you?

Demand

- What is the level of need for your service?
- Is the market for your service waxing or waning? Where are you on the "product life cycle" described in Chapter 4?
- Is there a *continuing* need for your service?
- What customers are entering/leaving this market?
- What consultants are entering/leaving this market?
- What future market developments can render your product (service) obsolete?

Prices

- Are prices (i.e., charge rates) stable, rising, or falling?
- Has the cost of doing business changed?
- Are competitors offering special pricing deals?
- How do your charge rates compare with your competitors'?

New Trends

- What recent developments offer new opportunities to explore?
- What new trends will affect your marketability?
- Will changes in regulations/standards impact your market?
- What new threats challenge your position?

Once you have answered these six sets of questions, you are ready for the next step.

POSITION YOURSELF IN THE MARKET

Now that you have a sense for the "lay of the land" and how the competition is deployed, you are ready to position yourself in the market. This happens in four steps:

1. Assess your strengths and weaknesses relative to the competition.

Inventory your strengths and weaknesses. Make a list of the technical "assets" that you have to offer a client. Make another list of your technical weaknesses and dislikes—liability areas that you want to avoid. Go into details; get specific. For each major competitor, make a similar list. If you don't have detailed information about some of your competitors, just use your best estimate of their strengths and weaknesses.

Now compare the lists: In what areas are you the strongest and they the weakest? These are the areas of greatest strategic potential for you. In what areas is the opposite true? These are the areas of greatest danger for you.

2. Define your target market segments.

To illustrate market segmentation, let's digress for a moment and talk about pizza! Pizza is one of the most highly segmented markets in the United States. The segmentation is based on:

- Size: personal size, family size, bigfoot size — enough for whole team
- Shape: round, square, etc.
- Price: inexpensive, affordable, luxury markets
- Thickness: thin crust, thick crust, deep dish
- Toppings: the usual mix, but always room for one more
- Delivery area
- Rapid delivery time, less than thirty minutes inside town lines
- Side orders available or bundled with — wings, celery, etc.
- Construction: How about cheese in the crust?
- Dough mix: flour, seasonings, etc.

All of the above constitute niches. Suppose you wanted to create a new niche for yourself. What new things could you come up with? How about "macho pizza" topped with beef jerky? Or how about "custom pizza" where customers specify their custom arrangement of toppings over the phone? How about "skinny pizza" with fewer calories and lower fat content? How about "kids' fun-time pizza" that comes packaged with small inexpensive plastic toys? How about a line of "chocolate lovers'" pizzas? Suppose your hobby is growing herbs. Can you come up with a "gourmet herbal pizza" that appeals to herb lovers?

By now you get the idea: in creating a niche, you capture a small portion of the market but forgo the rest of the market, where your competitors have *their* niches. That is, you stop trying to be all things to all people and offer what

- you are best in;
- gives you a natural advantage over the others.

By building up your clientele in your niche, you gain market share and stability.

OK, technical consulting isn't the same as selling pizzas — but the marketing principles are *exactly* the same! Suppose you are a mechanical engineer with a specialty in heat transfer. Believe it or not, some 5,000 people in the United States have experience and credentials in this narrow specialty. To create a niche, we must work a little harder. Let's say that your "heat transfer pizza" comes with computer chips on it. Automatically, the field is down to 500 people. Now let's say that you have worked long and hard to develop a computer program to analyze the transient thermal response of these computer chips. Suddenly, we're down to 100 people. Now let's say that you also have an extensive background in how these thermal transients affect the *reli-*

ability of the computer chips. Bingo! You have found a niche in which there are only 20 people in the country who can compete with your capability. Most of these 20 people are direct employees, so the number of competitors in your niche will be very small indeed.

This is an example of creating a market niche by *vertical expansion.* (See Figure 3.) It targets market segments that push deeper into what you're already doing. With vertical expansion, you not only stay within your area, you push deeper into one or more of its particular facets. You can also create a niche with *lateral expansion,* which increases the breadth of your current market mix. Instead of digging deeper in the same hole, you simultaneously explore different holes. Lateral expansion can occur in many dimensions:

- *Geography.* Increase the distance that you'll travel to gain clients. For convenience, you may have restricted your market to clients within a five-hundred-mile radius. Perhaps you should increase that radius or consider the whole country fair game. Some consultants explore worldwide markets to great advantage.
- *Technology.* Expand the range of services you offer into different, but synergistic, technical areas.
- *Business relationships.* Broaden your exposure to new opportunities by forming alliances, partnerships, ad hoc "teams," and reciprocal agreements.

EXPANDING YOUR MARKETS

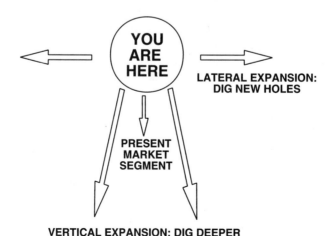

FIGURE 3. Market expansion.

- *Time or schedule.* Offer fast turnaround or round-the-clock service. For example, Dave is a computer hardware consultant whose customers can't afford downtime while their computers are being repaired. Many businesses are wary of repair shops that promise "tomorrow" but mean "next week." To provide an alternative, Dave works nights and weekends. He picks up the computers at the end of his customers' workday and delivers them before they start work the next day. Customers love Dave's service because they don't miss a single hour of productivity.

3. Create your competitive advantage.

This is where you get a chance to play the general and do some "strategic positioning." Once you understand your target market segment and your situation relative to the competition, it's not too difficult to find a spot of high ground where you hold the advantage. This spot of high ground utilizes your advantages at the very point where the competition is the weakest. There are numerous strategies you can employ to seize this spot.

- Wage a frontal marketing attack on the leader.
- Create an alliance for greater strength.
- Offer more responsive service.
- Create package deals that offer greater value to customers.
- Offer incentives for customers to switch over.

The choice of strategy also depends on your strength: if you're a small fish in a big pond, you may not have the resources to be "keeper of the gate." You may have to opt for "keeper of the garbage can" instead. As a general rule, breaking into an established market always puts the challenger at a disadvantage. In such cases, remember the old adage "If you're not the leader, never play by the rules the leader has set." For example, in the 1980s, LensCrafters challenged the established "rule" that you had to wait a week to get an eyeglasses prescription filled. It created a combination store-lab that could make eyeglasses in an hour.

As an example of how competitive edges work, I would like to relate my experience with Supreme Universal Corp. About ten years ago, I started consulting for them in heat transfer and soon became a regular. The projects I worked on were *very* complicated, requiring a mastery of aerodynamic heat transfer and numerical solution of differential equations. In addition, the problems required the ability to *quickly* piece together a practical solution by making and justifying technical approximations. Finally, these problems involved computer simulation and a degree of computer sophistication that

MEET THE PROS UP CLOSE AND PERSONAL

Deborah Kurata is principal consultant and cofounder of InStep Technologies, a California-based computer consulting firm. She is the author of *Doing Objects in Microsoft Visual Basic 5.0* and a popular seminar leader at technical conferences.

Deborah earned her B.S. in physics and went on to get an M.B.A. from the College of William and Mary. She has worked in the computer industry since 1980 and cofounded InStep Technologies in 1993. As a principal, she is responsible not only for getting her own work, but also for obtaining and managing projects for InStep's staff of sixteen subcontractors. In 1994 and 1995, Windows developers and consultants were relatively scarce, and marketing took little of Deborah's efforts. By 1997 an abundance of expert consultants were offering their services in the local market. Moreover, out-of-state consulting companies were opening branch offices in her local area. Suddenly, marketing was no longer as simple as it had been. To meet the competitive challenge, Deborah implemented a fourfold strategy:

First, and most important, she finds the very best developers available. She picks only staff who possess the required combination of solid experience and good attitude. To keep these folks motivated, she offers special benefits such as twice-monthly training sessions to keep them up-to-date on the latest developments.

Second, she keeps her name in the public eye through her books and seminars. Because writing a book requires expert skill and a huge time commitment, her competitors cannot easily duplicate this competitive advantage. Before Deborah was known publicly, she had to "prove" herself to clients by writing detailed proposals on her own time. Now that she is a well-known author, however, clients usually bypass the proposal phase and just write her a purchase order to come in. No more detailed and time-consuming proposals.

Third, she is careful to define her niche to be commensurate with the breadth and depth of the people resources in her company. Deborah says, "We don't want to focus so narrowly that new advances leave us without the correct skills. On the other hand, we don't want to focus so widely that we lose the ability to provide expertise."

Fourth, Deborah works hard on customer service. "Customers want and expect prompt, high-quality service, and we give it to them. If we have to go the extra mile, we do it. All our people are highly motivated extra milers."

was beyond the purview of most engineers working in the field. When you combined all these challenges, there were very few individuals—consultants or otherwise—who were capable of doing the job.

The fact that I happened to have this combination of skills, was, in essence, my competitive edge. I had spent literally thousands of hours of my own time studying these fields. (This was far less a burden than it might seem, because I had been fascinated by these subjects and had a great deal of intellectual curiosity about them.)

Since I had a competitive advantage, I was able to charge prices a little higher than my client was comfortable with. (That I was earning more than the division VP was not lost on any of the client's managers.) After six months, they set out to find a less costly replacement. First, they asked their personnel department to hire a direct employee to replace me. Over the course of the next year, they interviewed perhaps a dozen midlevel engineers. But when the client asked them how they would go about solving typical problems, none of the interviewees provided convincing answers or demonstrated a mastery of the field.

The client was interested in getting these problems solved *cost-effectively*. There were, by my estimate, about 500 people in the country who had the requisite combination of background and skills. Most of them were not midlevel engineers, but principal engineers who *already* commanded salaries that were 50 percent higher than the client was willing to offer. To offer them enough to accept a position would upset the client's entire salary structure. About 50 of these 500 were consultants. Yet, their charge rates were typically 30 percent higher than mine! So the client's challenge was to locate a competent consultant who, for some reason, had a lower hourly rate.

The client's next step was to hire a local college professor whom I'll call Nick. He had excellent credentials, and it was clear he had a mastery of heat transfer. Most important, because Nick was already earning a good salary at the university and was somewhat naive about business, he agreed to charge them about half my rate. The client gave Nick a consulting contract and set him loose on his first project. Nick worked diligently, but it was clear that he was not used to working closely with others. He stayed away from the client's office and never communicated his progress.

When the project was due, Nick delivered a neatly printed report with the solution. Supreme Universal was flabbergasted to discover that Nick had not understood the problem correctly in the first place. Nick provided a very elegant solution to the wrong problem! He had been asked to find the effect of pressure gradient on the heat transfer to a particular device (a "first-order" effect that could affect the answer by as much as 50 percent) and had instead delivered an analysis of the effect of viscosity on the pressure gradient (a "second-order" effect that affected the answer by less than 1 percent).

Yikes! When the client realized what Nick had done, they were crushed. They paid him off and let him go. After this incident, Supreme Universal finally stopped trying to replace me and gave me carte blanche for the duration of the project.

The reason I am telling you this is to reinforce the idea that after you have a few contracts with a given client, you can be sure that they will try to "optimize" their expenditures in your area. They will comparison shop for other ways to solve their problems more cost-effectively. Three things protect you, the successful consultant, against these efforts:

- Your competitive edge: specialized knowledge and skills that others don't possess.
- Your ability to solve problems cost-effectively: You may charge a higher hourly rate, but your solutions address the real problems and have lower *total cost.* Ultimately, the client learns that they can *rely* on you for these cost-effective solutions.
- An attitude of service and the discretion to *never* discuss finances with your client's troops.

4. Repackage yourself to highlight your competitive advantage.

Why do jewelers package high-priced watches and diamond rings in expensive boxes that will probably never be used again? Because it creates the *perception* of value in the customer's eye. Jewelers know that packaging is a very important part of the sale. A costly and elegant package creates the expectation that the watch or diamond ring inside is *also* valuable. In a similar manner, how you package yourself as a consultant has direct bearing on your sales effectiveness and on the prices you are able to command.

The way you package yourself as a consultant is very different from the way you have learned to package yourself as a direct employee. What are considered good "qualities" for workers in direct employment are not necessarily "virtues" for consultants. Therefore, consultants must repackage themselves in terms of

- defining their roles;
- highlighting what they have to offer;
- features that set them apart as consultants, not employees.

Repackaging turns "Me, me, me — look at my wonderful features" into "You, you, you — look at how I can solve *your* problem." That is, it describes and promotes your consulting service in terms of *what you can do for the client.* Repackaging doesn't change the contents (the skills you are marketing) but the outside (how you present it to the customer to improve its salability).

Always describe your previous work so that new clients can grasp the *value* of what you are selling. This is especially important for prospective consultants who have previously been forced into retirement or laid off, or who are having difficulty finding direct employment. Package the *very best* of what you have to offer and drop mention of whatever may distract or turn off a client. Focus on your exact market segment to convince clients that you are their best source for getting the work done cost-effectively. Here is how one software consultant repackaged himself to highlight his competitive advantage:

Original		**Repackaged**
I can do all languages.	⇒	Visual Basic and Oracle my specialty.
Whatever your industry, I can handle your project.	⇒	Bank managers: I know *your* business in depth and speak *your* language!
I have twelve degrees from M.I.T. I am great: Blah, blah, blah....	⇒	Look at what I did for these six banks and how I solved *their* problem.
I also have ten years' experience milking goats, just in case you need help in this area.	⇒	See my Web page for the latest information on how new banking rules will affect *your* data integrity system.

PUT IT ALL TOGETHER: YOUR MARKETING PLAN

We've come a long way in the marketing process. It's time to pull all the work together into a written marketing plan. The plan should spell out your marketing strategy and enumerate the tactics you intend to employ. At this stage, the planning process becomes more detailed and concrete. The five remaining steps are:

1. Target specific clients: in previous sections, we did this in terms of market segments. Here, I mean specific *companies.*
2. Determine the specific *individuals* at each company to contact. Reach out to them; let them know that you're available to help them with their problems. Determine specific promotion strategies to put you in better contact.

3. Create an action plan: a military campaign does not consist of a single foray into new territory and then waiting for things to happen. A general makes *repeated* and *coordinated* forays that constitute a *campaign.* Similarly, describe how *you* will make your second and third forays. Expect clients to receive you with some resistance initially. Also expect competitors to volley back; they will not stand by idly while you attack their position. Therefore, plan ways to counter client resistance and competitive threats *in advance.* Prepare new sales literature, demonstrations, and models to show your new capabilities. Define the unique twist you will add to give yourself sales leverage (your "unique selling proposition").

4. Spell out the resources required to put your plan into action. Create a budget for the needed cash outlays. Define specific tasks that place you within "striking distance" of the client. For example, before you talk to a client, you may want to review their annual report, study a relevant technical paper, or author a technical paper in their field.

5. Evaluate your plan in terms of feasibility and return on investment. What are the expected outcomes? How will you handle outcomes that *aren't* expected? Do you have a fall-back position and a contingency plan? Is your plan effective and efficient? Is it focused and complete? A quick feasibility check of your plan will pay handsome dividends.

How do you know whether your plan is solid *before* you implement it? You won't, but here are some reliable predictors. A good plan

- connects "here and now" with the desired goal in the future;
- is rational and realistic;
- makes explicit the resources needed and provides a timetable for action;
- is flexible and considers likely responses from others;
- is simple. Don't lose yourself by addressing a thousand "what if" scenarios! Use your common sense to focus on the few priorities that are really important.

SET YOUR PLAN INTO ACTION!

Some perfectionists feel that success only comes to those who "do the right things at the right time in the right place with the right people for the right reasons." This, however, is like saying that to win, all you have to do is beat everyone else. The statement is technically correct, but it sets an impossibly high standard for beginners.

My reaction to this attitude is that every consultant *grows* into his or her marketing abilities. At first, you will be doing many things that are nonoptimal (that is, the wrong place at the wrong time for the wrong reasons). This is inevitable. If you are like me, you have invested an enormous amount of time and energy in becoming a technical expert in your field. To insist that you suddenly become a master of marketing seems intimidating and unfair. I prefer a gentler approach: there are many marketing activities to be done. The most important task, though, is to start wherever you are. Get the ball rolling; you will learn as you go along.

**Get the ball rolling;
then incrementally optimize.**

Most people find that *deploying* their strategic plan is harder than writing it! Strategic plans often fail to be implemented because they require too many activities to be juggled at once. The resulting overload on time and resources leads to confusion and wasted motion. Hence, my repeated emphasis on keeping your plan simple and aiming only at your best shots.

After you get some results from your marketing forays, measure their effectiveness. Did you get the results you wanted? If not, modify your plans and efforts to get even better results in the future. As you grow professionally, you will develop a sense for experimenting with strategies to find the ones that give you the best results.

GUERRILLA MARKETING

So, you've put your marketing plan into motion and you're getting excellent results. Great! Keep up the good work! But what happens if your marketing plan is *not* working as expected? What are your next steps? Let's consider some contingency measures when your marketing results fall short of your objectives.

One of the primary reasons that marketing plans fizzle is that they fail to appreciate the fast pace and crowded conditions in today's marketplace. The pace of change is accelerating, especially in the high-technology areas that are fertile grounds for consulting. As these new technologies emerge, they compete for commercial acceptance, thereby threatening the old technologies. Some of the new technologies flourish, some merely survive, and others are quickly discarded like Sony's BetaMax and eight-track audio tapes.

The market fluctuates turbulently as this sifting takes place, making it difficult to be certain about which technologies you should become involved in.

In this new marketplace, other players are taking higher risks just to remain in the game. They are forming alliances to increase their leverage. Moreover, they have become more value conscious in marketing their customers. They know they must offer higher value and quality to their clients just to survive. In a marketplace this crowded, you must offer more *value* than you did last year—or other consultants will charm away your client base.

Older engineers and managers who are out of touch with these new market realities may offer you advice based on the "good old days." Such well-meaning individuals suggest that if you're having a tough time marketing, there's something wrong with *you!* They can cite example after example of *their* successes (in years past) and the strategies they used to achieve them. Remember, however, that in a crowded market, few people are inclined to admit how competitive things are. For some people, acknowledging a negative condition is tantamount to conceding defeat. Their philosophy is "Only losers and wimps complain. Successful people just try harder until they win."

I suggest a more compassionate view. When your market becomes crowded, it's tough for everybody. Look behind the facade companies put up. Industrial giants such as GE, GM, Ford, DEC, IBM, US Steel, Hughes, Boeing, and Sears (the list is much longer), with all their accumulated talent and skill, could not manage to turn a consistent profit in recent years. If your initial marketing plan falls short of its goals, you, too, should put up a facade. Crying on your client's shoulder will accomplish nothing. Maintain your dignity and composure. Then solve this marketing problem with compromise measures that get your cash register jingling again. It's time to meet the problem head on—with guerrilla marketing.

Guerrilla marketing is a set of low-cost, highly flexible strategies to successfully compete in a crowded marketplace. They depend on the individual consultant's ability to adapt rapidly to changing markets and to be more responsive to client needs. Guerrilla marketing demands greater resourcefulness, creativity, and aggressiveness than standard marketing techniques. Like a guerrilla warrior, you have to be willing to shoot from behind the trees, that is, deviate from the "rules," to give yourself an advantage. Here are some techniques that may work for you:

- *Try harder.* Pursue leads that previously failed or that were not immediately productive. Contact former clients. Like a guerrilla, target your clients and "stalk" them. Create strategies to attract their attention and follow up with attractive offers once you have it. In a crowded marketplace, significantly greater marketing effort is required. No matter how qualified you may be, *self-promotion* becomes the prime determinant of success.

- *Lower your standards.* Widen your range of clients and projects beyond what you would accept under normal market conditions. Some of the projects may be less interesting or savory, but they will keep your business alive.

- *Repackage your service.* Change your "product offering" to accommodate smaller companies and companies on a lower budget. Think of how you can put together a "turnkey" package that makes it easier for clients to use your services. Experiment with the way you describe your services in your brochure. Instead of a single general-purpose brochure, prepare three or four different ones targeted at specific market segments.

- *Diversify.* Experiment with services that complement your standard offering.

- *Lower your price.* Price slashing has dangerous consequences, so go easy on this strategy. If a new client says he can afford to pay you only 85 percent of your standard rate, take it. Explain that this is a special "get acquainted" offer, and that next year, your rate might be a little higher. Under no circumstances, though, do I recommend offering more than a 25 percent discount. (See Chapter 9.)

- *Form strong alliances.* Sometimes, an individual consultant must join forces to handle a large project. By forming alliances with other consultants and vendors, you can extend your marketing leverage.

- *Go where the money is.* Some of your former clients may fall on hard times. Even if they like your consulting service, they may no longer be able to afford it. Don't give up on the *individual*s; they still form part of your network. Instead, give up on the company. A guerrilla doesn't shoot at soldiers who are already dead. Likewise, don't waste time marketing a dying company.

- *Offer special promotions.* To attract new business, offer the client something related to your service that you can afford to give away. For example, if you consult on air-conditioning systems, perhaps you could advertise a free computer program that calculates frictional losses in ducts and piping. The only people who would be interested in such a program would be engineers who design air-conditioning systems. Voilà, the responses lead you to new clients.

- *Consider bartering.* Barter your prospecting tips with other consultants. You could ask for leads, commissions, reciprocal favors, or subcontracts. Make sure that the value received is mutual. Nobody will give valuable information time after time without wanting some form of consideration.

- *Wine 'em and dine 'em.* This classic marketing strategy works with some clients, although it can be quite costly. Research your prospects carefully to make sure they don't frown on this practice.

- *Create the problem.* Make your prospective clients aware of a new problem and then propose the means to solve it. To overcome resistance, play upon your clients' fears. I once witnessed a consultant inform a utility executive of a potential flow-vibration problem in his power plant. He showed the client a movie clip of the Tacoma Narrows bridge failure. (This was the most famous and disastrous failure attributable to flow vibrations. The clip showed the bridge literally falling apart.) At the end, the consultant exclaimed, "And your whole darn plant could fall down the same way!" With fear in his eyes, the client wrote a hundred-thousand-dollar contract on the spot to make sure that such a disaster would never happen to his plant.

- *Offer more flexible service.* If the competition is rigid in their service options, outsmart them by offering service that is faster, more responsive to the client's needs, more integrated with the client's equipment and procedures, and more in tune with the client's schedules, even if you must work nights and weekends.

Now, other consultants also know about guerrilla marketing. They, too, will engage in some unusual tactics. Expect the competition to "borrow" your ideas, methods, and promotions. In a competitive market, ethics flies out the window. Outright theft of intellectual property, especially by larger companies, is very common. They know that small companies and individuals do not have the time or money to exact justice in court.

There are two ways to counter guerrilla marketing of other consultants. First, *protect* your work. Don't make it easy for competitors to expropriate your unique ideas and methods. Second, even if your marketing promotion is successful this year, plan a new one for next year. Guerrilla warfare means moving from tree to tree to get the best shot. It requires more effort than traditional combat. Similarly, with guerrilla marketing, *always* stay one step ahead of the competition.

Marketing guru Robert J. Kriegel says, "If it ain't broke, break it!" In a book of the same title, he argues convincingly that in rapidly changing markets, there is only one way to ensure continuing success: Make your currently successful product obsolete—by bringing out its successor yourself. Don't rest on your laurels. If you don't "break" your own successful products, your competitors will do it for you!

Clients also respond differently when the market is crowded. They are flooded with aggressive marketers trying to sell equipment and services, so they tend to insulate themselves. It takes more work to meet and qualify them. The clients' needs still exist, but you must work harder to land the contract. Further, clients often react by placing unrealistic demands on consultants. For example, in the good old times, companies were able to address their technical problems in a timely fashion because development cycles were much longer. In newer, faster-paced markets, they let problems go unattended until the last minute and expect the consultant to "pull the project out of the fire." If the problem resolution takes two hundred hours of effort, they may ask you to fix it in eight hours—immediately. (Chapter 11 tells you how to overcome this obstacle.)

I wish that everyone could be a winner in this new economy, just as nearly every technical person throve in the booming fifties. But the stark reality is that in this new game, there will be winners and losers. Now that "showing up for the game" is not enough by itself, the solution is to become more focused, more efficient, and more strategically oriented in your marketing efforts.

INVITATION TO ENTREPRENEURIAL CREATIVITY

Consultants find it much easier to explore their own ideas than direct employees do. There is no limit to the creativity and innovation that can be brought into play. For example, you can market unsolicited proposals showing customers how to do something better, cheaper, or more efficiently. You can develop a new product with a client. You can create and market new products on your own.

As a consultant, you are your own business entity. This greatly extends the range of business options and arrangements that were available to you as a direct employee. Now you can expand your markets by using an agent or offering commissions and finder's fees for new clients. You can *create* marketing promotions that entice clients to come to you or give you leads for further work. Or you can act as an agent for other consultants in allied fields and obtain additional revenue in this manner.

Mike used creativity in marketing his services to MiniData, a small software company. MiniData had just published a program that did reliability analysis of printed circuit boards. Mike's specialty, it so happened, was reliability analysis of printed circuit boards! After reviewing the program, Mike saw three areas in which he could greatly enhance the program's capabilities. Of course, Mike knew that proposing a consulting contract to do this was unlikely to produce results. Software start-up companies are notoriously low on cash and high on "I'd rather do it myself."

The marketing challenge was to propose a deal that involved neither cash out-lay nor telling MiniData how to improve the program. Mike called the president of the four-person company and set up a meeting. After explaining his creden-tials and interest in the field, Mike offered MiniData a suite of proposals:

1. Could he buy their mailing list?
2. Could he review their software in technical journals in return for a list of companies on their mailing list?
3. Could he be their technical support "department" for technical problems they were too busy to handle?
4. Could he become a local rep to sell the software?

MiniData bit on the last two options. They were understaffed and unable to offer decent technical support over the phone. Using Mike as a "free" support office was appealing. It made them look better to their customers. In the process, Mike got to meet their customers.

The software company also liked the idea of Mike's becoming a local rep. They offered him $50 commission for each copy sold, a paltry amount. However, the real payoff for Mike in this deal was getting to meet clients with circuit board reliability problems. If the problems were serious enough to war-rant spending $500 on a program, Mike could convert some of these prospects into paying clients.

Over a six-month period, Mike gradually developed greater rapport with MiniData. They came to appreciate his expertise in devising algorithms for the complicated configurations. One year later, Mike negotiated a deal to write these algorithms for MiniData. Instead of billing for consulting time, Mike received royalties on MiniData's sales of the enhanced program.

The consultant can put together a *team* of consultants to bid on a multidisci-plinary contract. The organizer chooses his all-stars to do the work in *his* style.

Ernie was an engineering manager at a large optical design company before he became a consultant. Over the years, he worked with many consultants who supported the various specialties needed to produce an optical design. Through his network, Ernie learned of an opportunity to bid on a telescope project that was much larger than his consulting company could handle. Undaunted, Ernie decided to assemble a team capable of handling the project. He contacted the other consultants and asked each one to submit a résumé and help with their part of the proposal.

Ernie won the contract and then subcontracted with each consultant to get the job done. It worked beautifully. Each consultant—who was a known quantity by virtue of having previously worked for Ernie—handled his separate piece, and Ernie orchestrated the entire project. Ernie was still managing projects, but now he was the one getting the profits!

Creativity and resourcefulness allow you to overcome marketing obstacles that might stop others in their tracks. *Listen* to the client's objections and find alternate ways to circumvent the problem.

> Susan was having difficulty convincing her client that her proposed finite-element stress analysis would model the physical situation correctly. The client was skeptical. At the sales meeting, he jeered, "You know, just because you've drawn some fancy pictures on a computer screen doesn't mean that your model will accurately predict the stresses in my machinery. I want some assurance that you are not going to give me "GIGO" [garbage in–garbage out].
>
> Susan backpedaled her way out of the objection at the meeting but realized she would have to address the issue directly in her proposal. How could she provide the client the level of assurance he demanded? After musing about it for a day, she called the president of the best-known finite-element software company in the country. Susan asked if she could buy a day of his time, mentioning that she would be using his finite-element program to perform the work. (Susan estimated this would amount to $100,000 of program time, which provided additional incentive for the company president.)
>
> The president agreed. He would charge her only $1,000 for a three-hour technical review of her model. In her proposal, Susan mentioned that the country's foremost expert in finite-element analysis would review her finite-element model and provide critical feedback. The client's objections vaporized. Susan won the project. When she completed her model, she reviewed it with the finite-element guru and obtained his official blessing. The project was a resounding success.

I invite you to use your own *ingenuity* in marketing clients. Charles Thompson's book on stimulating business creativity (see the Suggested Reading List) offers a wealth of ideas to supplement your own.

HARVEY'S HANDY "HATE-TO-DO-MARKETING" CHECKLIST

New to this edition is Harvey's Handy "Hate-To-Do-Marketing" checklist for those of you who cringe at the thought of marketing. This checklist makes it easier to identify how long you can let certain market activities go before you must put on your marketer's hat and pretend you're working on your M.B.A.

Here is my list of the *minimum* recurring marketing activities to which you should be attending:

Yearly	❏	Conduct your one-day marketing retreat: Review last year's marketing plan. Did you meet your goals? What parts of the plan worked? What parts need to be revised? What new events or trends will affect your present strategies? Considering all these factors, create your next year's marketing plan, *in writing.*
Monthly	❏	Review your annual marketing plan: What parts must be implemented this month? Spend a few hours scanning for new opportunities and/or threats. Spend another few hours assembling a list of specific marketing tasks for this month. Assign each task a priority based on its potential payoff.
Weekly	❏	At the beginning of the week, take your monthly task list and pick the top three unfinished items. Write "assignments" to yourself into your calendar: name individuals to be contacted, identify letters to be written, designate specific reading or research to be done. If the tasks are too large to be completed in a week, break them into pieces than *can* be completed in that time frame. As you complete each task, check it off.
Daily	❏	Enter new contacts you have met into your database. As you come across it, enter market and competitor information into your filing system. Every day, review your marketing priority list and do your best to carry through on it.

Remember, if your marketing "department" doesn't follow through on its intentions, your company won't make it—no matter how good you are at your technical specialty! You will never have the fun of being your own boss unless you learn to apply just a little marketing elbow grease.

I hate marketing, but I hate the thought of professional subservience even more. Of course, it helps to state this in a more positive fashion. So, for the last ten years, I have placed a small affirmation card on my computer monitor, in constant view:

> ### Vote of Confidence Card
>
> Harvey has the intelligence, the ability, and the experience to make this business a success.
>
> **Now just DO IT!!**

CHAPTER 7

The Sales Meeting

Sales meetings with new customers are almost always challenging. To meet the challenge successfully requires adequate preparation and the ability to think on your feet. The experienced consultant knows that both of these requirements are developed through patient practice.

WHAT YOU MUST KNOW *BEFORE* VISITING A NEW CLIENT

Before you finish a telephone conversation in which you agree to go to the client's office, you should check that you have asked the following qualifying questions:

1. What is the specific need, both technically and projectwise? For example, determining the optimal cooling fan arrangement might be the technical need, and the project might be the client's forthcoming Model XX computer system.
2. What is the scope of the work? Does the client anticipate that it can be done in two days or in two months? Or is the scope currently undefined?
3. When does the client need the work done? Is it an immediate need for which they already have a contract, or is it for a proposal that has not yet come in?
4. How long will the meeting last? The consultant should suggest a definite meeting length if the client does not indicate a preference. For example, if a client asks me to meet at 1:00 P.M, I would tell her, "That is fine; it will give us plenty of time to talk before my next appointment at 3:15." The point is that you should not be available for an indefinite period of time, since the sales meeting is not billable time.
5. Are you the only consultant or company responding to the client's needs? How many others have been asked to propose?

Tip

When you go to the sales meeting, bring reports, photographs, and drawings that substantiate your background in your technical area. Also be sure to bring your standard sales material, including business cards, brochures, résumé, and other relevant materials.

Make all efforts to arrive on time for your sales meeting. Many managers have busy schedules and may not be able to accommodate tardiness on the part of the consultant. If you are late, the client may postpone the meeting, which gives your competition more time to market the client and make the sale.

Before you go to the sales meeting, look up recent articles and background material on both the specific technical subject and the project. An hour spent in this kind of research is a good investment, since it helps you present a more knowledgeable impression.

If the potential contract has a large dollar value, put proportionately more time into investigating the background. For example, suppose it becomes apparent that the client is talking about a $100,000 contract. Call the client a second time and get more details about the technical problem. For a contract of this size, you can afford to spend a few days preparing a technical presentation with visuals and handouts. You will need the extra detail to make sure you are on target with your presentation. Just showing up for the sales meeting without preparing is throwing away a good opportunity. You can be sure that your competition will treat the occasion more seriously. On your second phone call, establish with the client that you will be making a formal presentation and negotiate a suitable time and date for it.

WHAT CAN HAPPEN IF YOU DON'T "QUALIFY" A SALES MEETING

About twenty-five years ago, I worked for a small consulting company. I was their heat transfer and fluid mechanics expert. One day my boss walked into my office and said, "Pack your briefcase and put your coat on; we're taking a trip."

My boss had been called by an old friend at a Fortune 500 firm six hundred miles away. They had an extremely urgent problem involving flow-induced vibrations and wanted us to help. In his attempt to be "Johnny on the spot," my boss forgot to ask the qualifying questions outlined above. All he knew was that they were "hot to trot." There *appeared* to be good potential to land a large contract.

We flew to the client's facility without benefit of further details or purchase order. When we arrived at the client's offices, we were brought to a large conference room where the problem definition team was meeting.

Introductions were made. At the far side of the conference table, an old man was bent over a pile of papers. He was using an old-fashioned slide rule, and his face was pressed intently to his work. When he finally looked up, I couldn't believe my eyes. It was Yappi, the famous M.I.T. professor! We instantly recognized each other. Not only had I taken graduate courses from him, he had also been a colleague of my father's at M.I.T.

Yappi smiled at me. "Hi, Harvey, how are you doing? How's your family?" We started to chat. After a few minutes' conversation, the others drew closer to us. Now that we were all present, they wanted to start the meeting.

The meeting coordinator talked for about thirty minutes, during which time everybody listened attentively. Everybody, that is, except Yappi. Yappi hunched over his papers, working furiously. Suddenly, he stood up and stated, "Gentlemen, I think I have the solution to your problem."

Yappi proceeded to give an elegant and lucid description of the problem and described how it could be reduced to one of the cases in his textbook. In one hour of work, Yappi had solved the problem! We didn't have a chance against that kind of sheer brilliance.

I found out later that Yappi had consulted for the client on his usual terms, $2,000 (in 1998 dollars) for the day or any fraction thereof. We came out of the venture empty-handed.

There is nothing wrong with losing a chance to bid on a project. That's the nature of business. You win some and you lose some. There is also nothing wrong with my company sending two people on a *contingency* basis (that is, without a financial commitment from the client). But there *is* something wrong with expending effort in any situation where a few simple questions would determine that your chances of winning the job are minuscule.

WHY YOU WILL ALWAYS BE DOING SOME SALES WORK

Many consultants feel uncomfortable about the sales aspects of their jobs, even after ten or fifteen years of practice. Sales differs from engineering in content, demeanor, and attitude. It is unreasonable to expect an engineering consultant to be naturally attracted to sales activities when his foremost interest is engineering. On the other hand, I see no substitute for the sales meeting. Alternatives such as sales letters are generally a waste of time. The customer really wants to deal with you face-to-face, at least until he knows you.

Repeat business at a company eventually decreases over a period of years. There are a number of reasons for this trend. Your contacts at a specific company will leave, be promoted, or be transferred to different groups as time goes by. (Of course, this has the side benefit that your contacts migrate to other companies with which you are currently *not* established.) The company may no longer need your area of specialization. It may develop policies against hiring consultants except in dire emergencies. And some clients simply go out of business.

Therefore, do not expect to have a static list of customers who will always be there for you. It just doesn't work that way. There is no way to eliminate the sometimes uncomfortable necessity of sales meetings with new customers.

HOW TO DEVELOP SALES ABILITIES

Folklore says that some people don't need to study sales techniques, because they are "naturals." I wish to correct this mistaken idea: there is no such thing as a "natural salesperson." All of us—whizzes and fumblers alike—learn our sales techniques from a very early age.

If you were lucky, your early learning environment encouraged self-expression, confidence in front of strangers, and the knack of persuasion. With this background, you probably find sales fun and exciting. On the other hand, if your early learning environment made you shy, reserved, tentative, self-conscious, or fearful of talking with others, do not despair. The learning process never stops. Anyone, beginner or expert, can improve his or her sales abilities at any point in his or her life.

Sales abilities can be developed with a modest amount of effort, at least to the point where you are not making major mistakes. The techniques I am about to explain should give you a jump start. If you have no formal training in sales techniques, read *How to Master the Art of Selling* by Tom Hopkins or *The Feel of Success in Selling* by Jim Schneider. (See the Suggested Reading List.)

Selling your consulting services is easy when you follow the six simple principles on page 96 ("Sales in a Nutshell").

That's all there is to it! In the following sections, we shall look more closely at each principle.

SALES IN A NUTSHELL

1. Project a confident image.
2. Gain rapport.
3. Establish the need.
4. Show how you can satisfy the need.
5. Elicit the client's objections and concerns.
6. Lead—don't push—the client to agreement on action.

PROJECT A CONFIDENT IMAGE

Customers won't believe that you are offering a valuable service unless *you* believe it first. Therefore, the first step is to *believe in yourself.* Your inner attitude shows through. Your feelings color every word, gesture, and facial expression in every conversation you hold. Whether you are aware of it or not, others unconsciously pick up these minute signals you telegraph and sense whether you are happy being there, whether you are competent, and whether you are someone they want to do business with. They've got you pegged.

The only way to control these messages is from the inside out. This requires you to

- possess a sense of self-worth;
- understand that sales work is *part of your job;*
- realize that you are there to offer *value* to the customer. Your services are valuable because they solve the customer's problems.

Once you have this sense of self-esteem, you do not need to rely on tricks or high-pressure tactics to get contracts; you can simply use your ability to solve the client's problem.

Your beliefs are powerful forces that affect your actions. Figure 4 illustrates the links between beliefs (self-talk) and actions. What you tell yourself about your sales role affects how you *view* yourself in that capacity. This image, be it positive or negative, affects how you *feel* about the selling situation. If you view selling with fear or as a necessary evil, you will subconsciously undermine your own efforts. Feelings of inadequacy, self-betrayal, or anxiety will surface.

The cycle continues. Negative feelings in the sales meeting generate negative *actions.* The consultant may become withdrawn, judgmental, defensive,

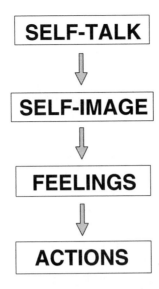

FIGURE 4. The belief cycle.

aggressive, or solicitous. Depending on the client's response, the meeting can degenerate into hostility, awkwardness, mutual noncommitment, and other no-win scenarios.

To become more successful in selling your consulting, work on the beginning of the belief cycle. Change your heart and your beliefs about the value of the sales activity. Sales allows you to display your wares to your customers. You may be the most brilliant scientist or engineer who ever existed, but if you cannot convince clients that you can solve *their* technical problem, nothing will come of your brilliance.

Selling is more than a necessary evil. It is the first step of performing your service. It requires attention and a positive attitude to transform the sales activity from a chore into a key activity. With a positive attitude, your consulting practice will grow and prosper. You will present yourself with dignity and will convey a sense of purpose that commands respect from clients. Coming from this positive mind-set, you will be able to treat all clients with the dignity they deserve. Believe in yourself and your ability to serve the customer.

Your belief in yourself must be strong enough to resist opposition.[1] The acid test of the strength of your belief comes when prospective clients curtly reject your proposals, laugh at your ideas, or treat you shabbily.

---

[1] You don't have to be stuck with the belief system you acquired over the years. With conscious effort, you can change it. In *Awaken the Giant Within,* Anthony Robbins gives some excellent techniques for changing limiting beliefs and getting the success you desire. See the Suggested Reading List.

Tip

If you have mixed feelings about an upcoming sales meeting, perhaps you do not have enough *information* about the client's needs. Going to a sales meeting without a good idea of the client's interests is like going to the grocery store without bringing your wallet. Every motion you make will ultimately be wasted. Therefore, interpret such feelings of uneasiness as a sign that you need to do more research or ask more questions.

Fear of rejection is a major barrier for anyone doing sales work. The trick to overcoming it is to learn to *separate the result of the sales meeting from your self-worth.* Expect many noes before you get to the yeses that will skyrocket your business to success. Most clients have a heart; they usually buffer their noes with compassion and kindness. When you come across the occasional grump who wants to ventilate his anger by attacking you personally, let the insult go. It's his problem, not yours.

GAIN RAPPORT

Generating rapport is a skill that can be learned. The goal of rapport is to break down the barriers that prevent straightforward communication. You will know you are successful when you establish a *dialogue* that is balanced and flowing.

The first part of gaining rapport happens before you ever set foot in the client's office. It is *learning the client's "language."* Become familiar with the client's business and operations. Learn the buzzwords particular to the industries you are marketing. Without this knowledge, you risk appearing like a dunce at sales meetings. Worse, you could appear insensitive to important company issues. It is easy to garner this information from journal articles, advertisements, the company's brochure, and the company's annual stockholder's report (if the company is publicly held).

The second part of gaining rapport happens before you ever open your mouth. Your appearance should fit in with the client's corporate culture. In Chapter 10, you will see how to dress and act in a manner that meshes smoothly with the client's expectations. Even possessions such as your car and briefcase make a statement about you that must be reasonably congruent. If you choose to clash with your clients' corporate images, you will make them uncomfortable and apprehensive. ("Why is this consultant wearing a tuxedo to my machine shop?")

The third part of gaining rapport is establishing the dialogue. Clients are no different from other people. Until they "know" you, they will be reluctant to open up. Use your conversational skills to *establish points of common concern*. When you attain rapport, the quality of the conversation changes. It is no longer two monologues with "your interests" and "my interests," but a dialogue with "our interests."

Overcome the client's resistance to openness by first talking about neutral and public subjects. (The weather!) Gradually work into more personal subjects that indicate a *common* interest or viewpoint. This is the tricky part. Avoid controversial topics such as politics, sex, and religion. Aim for popular subjects such as sports, the company's background, or highlights particular to the client's location. ("Everybody here mentions Cajun cooking. What's that?")

You might comment on something about the client that catches your eye, perhaps an interesting picture hanging on his wall or a model of the client's latest project. I sometimes use national backgrounds to segue into my interest in philately. For thirty years I have collected postage stamps from around the world. When I meet a client with a foreign name or accent, I discreetly establish his nationality and mention that I enthusiastically collect stamps from his country. If I get positive feedback with this approach, I go on to briefly describe one or two of my favorite stamps from his country. This rapport-generating trick works about 80 percent of the time. It depends on my extensive knowledge of philately. Use rapport generators that draw on *your* individual interests and strengths, be they football, golf, sailing, photography, or whatever.

How do you know when you are achieving rapport? Look for signals of receptivity. Some of these are direct eye contact, smiling, nodding, relaxed posture, absence of facial muscle tension (such as squinting the eyes), and leaning toward you.

Some signs of losing rapport ("resistance") are crossed arms, jittery fingers, foot tapping, easy distraction from outside noises, eagerness to take phone calls, frowning, pushing away from you, eyes not looking straight at you, tight lips, and fingers touching the face or covering the mouth. Glancing at the clock frequently, playing with jewelry, and hair pulling are other telltale signs that you are losing rapport.

Twisting rings on the finger is a sign that the person is nervous about what

that ring represents. If it is a wedding ring, it means that something in the conversation reflects on his or her marital situation. (Under these circumstances, never tell a husband-wife joke.) If he or she is twisting a school ring, it might reflect concern about his or her technical abilities, especially if you are wearing your class ring. When you observe ring twisting, it is a sign to change your own behavior, to stop pushing the client's unconscious "button."

Nonverbal language applies in client meetings with persons of the opposite sex. Suppose a male consultant meets with a female client. If the woman covers her blouse with her hand, it usually means one of two things: either the man is staring where he shouldn't be staring, or he is pressing too forcefully in the sales presentation. In this situation, the man should back off a little—physically and saleswise.

There are parallel concerns when a female consultant meets with a male client. The woman must decipher the male client's nonverbal behavior to understand whether the client is being receptive to her technical presentation or to her personal self and modify the sales strategy accordingly.

You can also use nonverbal information to better assess the client's *style*. This "meta-information" will help you decide how to best sell your consulting service to that person. To understand the client's style, ask yourself:

- Is the client talking mostly about people and companies, or about products and concepts?
- Is the client asking for details or sweeping assessments?
- Is the conversation balanced or is the client controlling it inordinately?
- Is the client easily distracted? Is there an excess of small talk after rapport has already been established? Does the client want to get right down to business?
- How confident is the client?
- Is the client treating you respectfully?
- Is the client rigid or flexible in exploring the issues?
- Is the client straightforward? Do you suspect that information is being withheld?
- Is the client a doer, a complainer, an observer, or a leader?
- Is the client enthusiastic about your presence?
- What do the client's clothing and office decor tell you about the client?
- How does the client treat co-workers who are present at the meeting?

Once you know the client's style, you can structure your responses to better achieve rapport. If you are dealing with a *thinker,* for example, emphasize the rational reasons and benefits. Take your time to explain the reasons for your questions and suggestions. Don't use sales gimmicks or appeals to emotion. And above all, don't use high-pressure tactics.

In dealing with a NIHer ("Not Invented Here" type), everything you say will be twisted to prove that the client's approach is the only one that is reasonable and practical. The best response is to treat such a person with great respect. Make him or her feel important. Then ask questions that compliment the person on the one hand, but that also reveal project objectives currently not being met.

In dealing with a *risk avoider,* explain your credentials and offer references to other projects you have successfully completed. Anticipate the client's excessive concern with loss by showing how you can provide a risk-free solution. Offer guarantees and low-risk means to start doing business together.

The ultimate goal of rapport is to achieve mutual trust. You must sell yourself before you can sell your consulting service. You must appear as reliable, competent, cooperative, and a fair player before clients will consider playing ball with you. Creating trust can't be forced. You can't trick anyone into trusting you and get away with it for very long. Therefore, trust depends on the sincerity of your intentions. Are you there to help the client and offer value for your fees? Or are you there to power play the client into buying something he doesn't need?

> **YOU MUST SELL YOURSELF BEFORE YOU CAN SELL YOUR CONSULTING SERVICE.**

Part of this sincerity is demonstrated by the way you conduct yourself during the sales meeting. You must genuinely listen to the customer so that you can understand the situation. The client can tell whether your attitude is empathetic or manipulative. You must explain your purposes so that the client knows what your intentions are. In this regard, establishing outside references also increases your credibility.

When people don't trust each other, they hide their priorities. It becomes unlikely that they can meet each other's needs this way. Instead, they display false agendas to each other and withhold information that could open up useful alternatives. When there is lack of trust, each party is so intent on protecting himself from manipulation that the number of mutually satisfactory outcomes is severely limited. Hypervigilance for the other party's hidden motives hampers cooperation. The sales meeting then becomes a contest in which each side strategizes to gain the position of power.

ESTABLISH THE NEED

Once you have established rapport, you are ready to understand the client's needs. In this phase of the sales meeting, keep your eye on the customer and your thoughts on the customer's problem! If you start to think about your own problems, you'll distract yourself and lose the ability to think on your feet.

The conversation should center on the client's problem. Don't start talking about your own background first! There is a good reason for this: you will not know which aspects of your own background to present in support of your ability to solve the client's problem until you know *what* the client's problem is. Simple? Yes, but perhaps 50 percent of all consultant sales meetings have negative outcomes because of this incorrect focus.

The major tool in establishing the client's need is *asking questions.* The consultant uses skill and savoir faire in probing the issues and getting to their core.

Early in the meeting, try to determine the client's motivations for buying your services. Is it to fulfill a single pressing problem? Is it to replace an existing consultant with whom he is not satisfied? Are you providing another bid on a government job that requires three bids to assure competitively low prices?

Make every effort to be nonjudgmental in asking your questions. If you don't keep your prejudices to yourself, you run the risk of self-sabotage:

Consultant: "You're not going to use the Star Trek model 6 tricorder, are you? It's a loser!"

Client: "Well, our president insists it's the best unit on the market right now, and I agree with him."

Consultant: (Oops!)

 Tip

Start with open-ended questions. In the beginning, avoid questions that can be answered with a simple "yes," "maybe," or "no." Open-ended questions get the customer talking. (For example, "What do you think of...?" "In what way is...a problem for you?" "What would be the ideal solution to this problem for you?")

Tip

One trick I use to remember this in the middle of meetings is to think, "The person facing me is the smartest person in the world, except for my one small specialty. Let me treat this person with great respect."

When you are asking questions about sensitive matters, exercise tact and diplomacy. Phrase questions that are necessary but potentially embarrassing so as to provide *graceful egress* for the client. Discreetly avoid questions that sound like you are trying to determine guilt or assign blame. For example, suppose someone in the client's company has botched a project. You have been invited to fix it. In phrasing your questions, tiptoe around the issue that a screwup has occurred. Never refer to the guilty parties as incompetent or technically inept.

In probing the client's need, don't make your questions too difficult to answer. Many consultants ask convoluted questions as a form of technical one-upmanship to demonstrate their technical superiority. It never works. Truly superior technical consultants gauge the depth of the client's knowledge and phrase their questions accordingly.

If you ask questions for which the client is not prepared, preface them with an explanation. For example, "Before I can determine the cooling system that would be best for you, I must know whether the unit will be operating inside, in a controlled thermal environment, or outside, exposed to summer and winter temperature extremes."

THE MAGIC POWER OF LISTENING

One important aspect of asking questions is often taken for granted: listening to the answers. You can help your clients feel comfortable during the fact-finding phase by being a good listener. Even though you are on the client's turf, it is the *consultant* who must consciously enable the conversation to flow. Think of the transitions you must make in your conversation to bring the client gracefully to the various subjects you are probing. If the client rambles in a different direction, don't get flustered. Hear him out. Then ask another question that bridges the immediate topic with one that puts you back on track.

LISTENING SKILLS

- Listen with an open mind. Don't let your biases filter out important information.
- Give the person your undivided attention. Use your whole body in actively listening. Look directly at the client while he or she is talking. Nod at appropriate points to provide feedback.
- Don't guess what the client means on important points that are unclear to you. Politely ask for clarification.
- Don't interrupt the client. Even if you know the subject a hundred times better, interrupting is rude. It is sure to turn the client off. Interrupting is an implicit message that you *don't* want to listen.
- Listen for vocal inflections. When the client emphasizes a word (e.g., "That was a *bear* of a problem for us"), it is often a direct invitation for you to dig deeper along that vein.
- Pay attention to the client's body language. If rapport is already established, fine. But if the client is glancing at her wristwatch or the clock every few minutes, it means you are pushing the limits of her patience. Listening to this nonverbal language allows you to spontaneously modify your sales approach to accommodate the situation.

Listening is an art. It requires paying great attention to the elements of the client's problem description. Then you must weave these details into your next question to maintain the direction of the conversation. Your job is not merely to keep the ball rolling, but also to lead it to the vicinity of the goal. Intelligently structuring and sequencing your questions—and then listening carefully—allows you to do exactly this. When you truly listen to the client, it is much easier to build a convincing sales pitch. In his book, *Niche Selling,* William T. Brooks observes, "You cannot talk people into buying, but you can listen them into it."

SHOW HOW YOU CAN SATISFY THE NEED

The techniques of sales persuasion can be summarized in a single sentence: Convince the customer that you are not selling *your* product, but fulfilling *his* needs. For example, AT&T doesn't advertise, "Make lots of phone calls" (that is, use its service.) Instead, it invites you to "reach out and touch someone" (that is, fulfill your needs).

For the technical consultant, this idea translates into: You are not offering technical services, but *solving your clients' critical problems.* Your technical capabilities, however excellent, are not what will convince the client to give you a contract. Rather, it is the idea that your expertise and equipment and computer programs can save the client's day and provide a cost-effective means of addressing the problem.

Engineers are notorious for trying to sell a product based on its features, features that they themselves may have designed into the product. It takes a long time for technical people to realize that customers do not buy a product or service based on its features but rather on the *perceived benefits.* A famous sales master, Elmer Wheeler, said it sixty years ago: "Sell the sizzle instead of the steak."

What does this mean to a consultant selling technical services? Well, for one thing, don't try to convince the client of your expertise or technical capabilities. Instead, focus on the client's problem and explore the ways you can solve it. That is, the "steak" is your technical expertise, but the "sizzle" is solving the client's problem.

Features include your special capabilities, the special equipment you use, or the computer programs you have written. Features are what allow you to offer the benefits to your clients. The sales challenge is to *translate* your features into the benefits that the customer will receive. Phrase them in the very language that your client uses.

For example, if a client is interested in having his toaster pass a certain UL® test, the one-dimensional transient thermal analysis I might propose is just the steak. By itself, this analysis will not strongly motivate the client to buy. The client doesn't care one whit whether I use one-dimensional or three-dimensional analysis. The client is likewise not concerned with whether I use analytical methods or a finite-element computer model. All the client cares about is: "Gotta pass the UL® test." Under these circumstances, I would try to sell an Underwriters Laboratories[2] Thermal Test Plan.

For technical consultants, the most important part of this phase of the sales meeting is scoping out your *technical approach.* Outline the methods you will use to solve the problem. In doing so, you are simultaneously creating a preliminary *statement of work.* That is, each step that you verbally outline will later become a "statement of work" item in your proposal. (Proposals will be discussed in Chapter 8.)

---

[2]Founded in 1894, the Underwriters Laboratories Corporation is an independent nonprofit corporation with a staff of thousands. They set standards for commercial products in which safety is a concern such as computers, space heaters, and household appliances.

For each specific task in this "shopping list," *briefly* describe how you will go about accomplishing it. Don't go into details unless the client insists. Convey the impression that you can easily handle the technical demands of the project. Look for nods and other signs of agreement from the client. When a client frowns or looks blank, he or she probably has reservations about the immediate topic. This is a signal to ask more questions to get feedback.

> *Consultant:* "You don't seem very receptive about my determining the thermal properties of silicon in this phase of the project. Is there a problem with that?"
>
> *Client:* "Well, we have a company policy. All technical work must use our tabulated material properties for the sake of consistency."
>
> *Consultant:* "Fine, I'd be glad to use your property tables in my analysis."

As you get further into the sales meeting, think of good reasons your service will offer more *value* than that of your competitors. Since value is in the eye of the beholder, you must interpret it *for each client in each situation.* Show the client how your service or solution is

- More reliable
- Faster
- Better able to take advantage of the latest technology
- More efficient
- Better able to integrate into the overall product design
- Better at technical support
- More responsive to client needs
- Most cost-effective
- More convenient
- More credible

Spell out these benefits! It is a mistake to assume that your clients can always discern the benefits of your services for themselves. Maybe you think such things should be obvious, based on your impressive credentials, multitudinous accomplishments, or innovative products. If so, you are guilty of looking at the sale from *your* viewpoint instead of the customer's. This is no occasion for conservative understatement or professional reserve. Enthusiastically state the benefits the customer will receive as a result of using your service.

Be prepared to *prove* that your service offers the benefits you claim it does. Show the client journal articles, photographs of completed projects,

> **SPELL OUT THE BENEFITS! DON'T ASSUME THAT
> CLIENTS CAN ALWAYS DISCERN THE BENEFITS OF
> YOUR SERVICES FOR THEMSELVES.**

design documents, brochures, and letters of referral from associates to sub-stantiate your claims.

Being persuasive is like judo. Observe how the other person is situated and use his momentum to move him in a direction you choose. In this case, the momentum is his own motivation and desire. Learning to sell successfully is learning how to sell from the client's viewpoint. Discover how to use his wants and needs to get what *you* want out of the interaction.

This client-centered approach means that there will be times when you should give up on selling a particular client for that day. You shouldn't talk people into anything that violates their best interests. It is better to walk away empty-handed than to be remembered as "the silver-tongued fox who sold us Eskimos an air conditioner."

Expect to do a certain amount of thinking on your feet in sales meetings. Don't try to plan everything in advance; allow for spontaneity. If you have pre-pared a "canned presentation," don't feel that you must rigidly adhere to it. One problem with canned presentations is that you start talking before you know whether your comments will be on target. To minimize the potential damage, brief the client contact person *before* the meeting. Make sure you have covered the bases and sidestepped incendiary issues.

In some sales situations, an important issue is convincing the customer that you can quickly come up to speed on details that are new to you. By and large, clients will not pay you to learn anything on their time. Therefore, you must sometimes give the impression that you already know the technical issue at hand and later do some "speed learning" to cover your bluff. Success in consulting depends on developing a feel for just how much you can extend yourself in this manner without getting into trouble.[3]

As a result of the information explosion, all experts today have large gaps in their knowledge. When a client has a complex technical problem, no sin-gle individual may be familiar with *all* of its technical ramifications. That's why big-league consulting companies like Arthur D. Little, Inc. maintain large staffs. They eliminate client objections that the individual consultant handling the project doesn't "know it all." On the other hand, many clients

---

[3] At other places in this book, I recommend never bluffing on specific technical points. Here, the bluffing merely asserts your general familiarity with a subject and is less likely to damage your credibility.

Tip

Be flexible with regard to the client's time. Some clients may be called out of the meeting before your time is up. You and they may have only thirty seconds in which to decide whether to condense your pitch or postpone the sales meeting. In such a situation, I always opt to condense. I ask for ten minutes in which to give the highlights—and then hit them with my best shots. If they are interested, they usually request another meeting to continue. Choosing to postpone straightaway, on the other hand, gives your competitors the opportunity to win the contract before you even get a chance to open your mouth.

discover that the costs of having this wealth of capability on tap are enormous. I have seen big-league consulting companies charge ten times more than an individual consultant for the same work. Large consulting companies must recoup extremely high overhead and marketing costs.

Most clients understand these trade-offs. They are willing to give a credible individual consultant the benefit of the doubt. The door is open to come up to speed in the few aspects of the project that you don't already know. You don't have to become an expert in these peripheral areas, only sufficiently adept to do the project.

For example, I once received a contract to do a finite-element thermal stress analysis of a mechanical assembly. Both heat transfer and stress analysis are within the purview of my expertise, and I felt comfortable. As the project progressed, it became clear that the temperatures were high enough to cause plastic deformation. Alas, plasticity was not my strong suit. Nevertheless, I devoted three weekends of study to it and successfully handled that part of the project.

ELICIT THE CLIENT'S OBJECTIONS AND CONCERNS

The previous section showed how judo techniques can be used to turn the client's needs into advancing your sale. You can also use judo techniques to handle client objections. Whenever a client poses a strong objection, it is an opportunity to bring him closer to the sale. In judo you do this by giving in to the opponent's force and throwing him in the direction in which he is moving. In the selling situation, the client's objections are used in a similar way. Instead of resisting the client's objections, ask to understand them in greater depth. Don't defend yourself against the objections—explore the objections. By going deeper into the objections, you are defusing the client's emotional reactions. In the process, you will learn enough detail to structure a response that emphasizes your strengths.

The language you use in this judo is important. Don't *agree* with the client's objection, but seek to *understand* it. For example, "I can understand why maintaining low hourly labor rates is important for you. Many of my clients have the same concern. However, I can show you how many of them were able to *save* money on the overall project by using my unique thermal analysis computer program. Here is a letter from one of my clients, thanking me for the $12,000 my method saved them compared with the old-fashioned way."

In Chapter 6, we saw how Susan sold a consulting project to a reluctant client. Instead of giving up, she used the client's inordinate concern for high-credibility assurances to advantage. ("Young lady, how do I know that your solution will not be GIGO? I want an expert with a national reputation to assure me that your answers are correct.") Had the client not thrust his arm out this way, Susan would have had little with which to "throw" her opponent. She overcame the client's objections by first listening to them and then responding to them in a way that would convince the most ardent skeptic. ("You want a super expert? OK, I'll get the number-one expert in the country to review and bless the analysis!")

Don't react to objections by *arguing* why they are unimportant or not valid. Never tell the client, "You're dead wrong!" Even if the client *is* dead wrong, he will resent your assessment. You may win the argument, but you will certainly lose the contract.

Make every effort to prevent objections from turning the meeting into a verbal boxing match. Don't get flustered. Maintain a friendly attitude. To give yourself more time to think of appropriate responses, repeat the objection to the client in your own words and ask him to be even more specific ("You say that C++ is not the correct computer language to use for this project. In what respects do you feel it is inappropriate?").

Answer objections by first softening the blow ("I see how you could feel that way"). Then allow the client to save face ("Many others have felt the same way, and with justification"). The next step is the critical one. You assert, "But I can give you three good reasons it is to *your great advantage* to consider another viewpoint for a moment." This is the bait: There are benefits of which the client may not yet be aware that outweigh the objections. Finally, explain how these benefits offer greater value. This is also the perfect time to pull out testimonial letters from other clients to strengthen your case.

Always carry extra materials that can help you overcome objections. For example: "I can see why you view my inexperience in high-temperature molecular dissociation as a shortcoming. It's true that I haven't written a journal

Objection	Rebuttal
Our need is not immediate.	Things take time; if we start the paperwork now, it will be ready for project kickoff.
Your charge rate is too high.	Even though I charge more, I can show you how my service is more cost-effective.
We have already retained another consultant.	I can do certain *parts* of the project faster and better. Let me provide support in these areas only.
You lack experience with our special application.	It will take me only a few hours to come up to speed. I can even do the background reading at home on my own time.
You haven't been in business long enough to be a "known quantity."	I can furnish you with three good references from companies where I completed similar projects as a direct employee.
Your experience has been mostly in another industry, not in ours.	The techniques are very similar. Here, look at these examples and I'll show you how.
We're too busy to consider your proposal.	All I ask is that you spend five minutes to consider a proposition that could save you many thousands of dollars.
We don't hire individual consultants as a matter of policy.	I offer a unique service that is unavailable from your standard consulting vendors.
We don't agree with your technical approach to the problem.	Let me explain to you why I think it has merit and why the other approach may lead to trouble.
You don't have sufficient credentials.	Perhaps, but these examples prove that I can do this particular task quickly and competently.
My boss has to approve it.	I appreciate that fact. Let's schedule a very short executive summary meeting for her benefit.
We can get another consultant who charges less.	True, but will the other person be able to do a credible job in the same short time frame?
We doubt your ability to get the work done in this time frame.	I will give you monthly/weekly/daily progress reports. Also, let's negotiate a penalty fee in case I fail to deliver by the project due date.

article on this subject. But let me show you a report here in my briefcase. It's from a project I did last year on high-temperature ionization. The calculation methods and thermodynamic principles are very similar."

Handling objections with diplomacy and finesse takes practice. It's worthwhile rehearsing your response *before* the sales meeting. To get you started, here is a list of common objections and possible ways to overcome them.

LEAD—DON'T PUSH—THE CLIENT TO AGREEMENT ON ACTION

Now that you've shown the client how you can solve her problem, it's time to wrap things up. This part of the sales conference is played by ear. In 90 percent of all sales meetings, by this time you will know whether the client is interested in giving you a contract.

Closing the sale takes the client from "interest" to "making a commitment." As the salesperson, you have the job of helping the customer make this leap. Toward the end of the meeting, start dropping hints about including certain items in your proposal. If the client nods, you know that he expects you to make a proposal.

In situations where the client's interest level has been high and the sales meeting has progressed well, close by simply asking for the client's business: "I'm very enthusiastic about working with you on this project. What is our next step? What paperwork do we need to make this official?"

If they say you have the contract, fine. If they refuse you flat out, that is also fine. When a client categorically denies interest, at least you know not to spend further effort marketing them. Many responses fall into the gray area in between. Some clients balk or want to "think" about it. If you and the client are still unclear about the course of action you should take, consider these options:

- You "think" about it also. Maybe your technical approach could be changed to better address the problem. Offer to go back to your office, consider the problem in greater depth, and come up with a more suitable approach.
- Ask for more details about the client's exact needs. Then suggest making another presentation that uses these details to provide a custom solution. The action item here is gaining consensus to come back for another meeting.
- Often the client is unsure because he is considering other options at the same time. Agreeing to wait is OK, as long as you find excuses to call the client later and monitor the status.

- If the client hesitates because he doesn't know when the project will start, convince him that there is no risk in writing you a contract *now*. Offer assurances that you won't start work on the contract until the client "turns it on."
- If the client remains unsure of your ability to handle the project, offer to send the telephone numbers of references who can attest to your technical capabilities.

Pave the way for following up after the meeting. Counter "Don't call us, we'll call you" responses with suggestions that keep you in the information loop. This is usually easy, since most sales meetings unearth numerous technical issues that serve as an excuse to call the client back.

Do not use high-pressure tactics to close the sale. That is, don't push the customer! *Lead* them to making the commitment to buy. Convince them that it's in their best interest. Selling consulting is not like selling televisions. Hard sells and last-minute desperation moves are unprofessional ("Wait! I can give it to you for wholesale" or "Sign me up now or I may be busy when you need me").

"Closing" means getting a commitment from the client to give you a consulting contract. Usually, this commitment is first verbal and then followed with the formal contract. In dealing with small companies, you may get a yes right on the spot. With large companies, cutting a contract requires many signatures in the approval cycle. Some managers in large companies are wary of voicing commitment until the approval cycle is complete. Therefore, with large companies, do not expect an instantaneous and official yes. Instead, aim for informal commitments and, most important, a request to submit your proposal or quotation letter. Chapter 8 shows you how to respond once this request has been made.

SOLVING YOUR CUSTOMER'S *REAL* PROBLEM CAN PAY OFF

About ten years ago, I was helping a client develop a thermal protection system for an electric actuator in a high-performance missile. As we got further into the project, it became clear that we would need to determine the effects of ablation on the actuator's performance. (*Ablation* is the high-temperature thermal erosion of the missile's surface.)

One of my client's senior managers called the largest and most prestigious consulting company that specialized in ablation technology. He briefed them on the problem and asked them to propose a method for assessing these effects.

Two weeks later, the large consulting company sent a proposal team to my client's offices. Two of the three team members were nationally known experts in the field of ablation. The third was a marketing manager who was polished, professional, and totally charming. We had a daylong meeting, during which the consulting company gave an exhaustive presentation of the many sophisticated computer programs they had developed to handle ablation phenomena. Their presentation was smooth and well rehearsed. At the end of the day, they handed out a beautifully prepared fifty-page proposal.

Having already worked on the project, I knew that the answers from a sophisticated ablation program would not solve the *real* problem. My client thought the whole problem was simply to determine the ablation process. This was a misperception. The real problem was: How did the ablation affect the *actuator?* That is, how were the two coupled together? I knew that this coupled approach could be accomplished by making a few simplifying and justifiable assumptions. The large consulting company seemed uninterested in determining whether such simplifying assumptions could be made. They made no effort to find out. The large consulting company did not want to solve the client's real problem, but only to sell what they were interested in: a program of research and development concerning ablation.

I felt the large consulting company had missed the boat in defining the problem, so during the next week I put together my approach to the problem. I prepared a number of viewgraphs that showed how their approach would give an exact answer to the ablation question but would not address the real issue. I proposed a simple method to get the answer to the real, combined problem.

I had sufficient credibility with that client to ask permission to present a one-hour slide show. Before my presentation I hand delivered a one-page meeting outline to each manager. As I gave them the sheet, I said that my alternative to the consulting company's approach would save $150,000 and five months' time. Claims like that generate interest!

At the meeting, I explained the real problem in language chosen for maximum clarity. Then I described and compared the options available to solve the problem. I finished with my recommendations and plan of action. I anticipated major objections by preparing additional slides in advance. By the end of the meeting, I could see that I had generated total rapport. As the top manager walked out of the conference room, he told me in his unique style, "It's all yours, baby."

Perceiving your customer's real needs can give you an advantage over even the most credible competitors. In selling your services, therefore, *listen with your brain running.* Has the client perceived and stated the problem correctly? If not, can you propose a way to better address it?

MY "SECRET WEAPON" MAKES CLIENTS SALIVATE WITH INTEREST

When I go to a client's office for the sales visit, I bring along my "secret weapon." This is nothing more than an attractive portfolio of my previous projects. The portfolio is in a leather-bound three-ring binder that contains a number of transparent polystyrene sheet protectors. Each sheet protector encloses a sketch or a photo with a few words describing a particular project. These are arranged so that the idea of the project can be grasped in a single glance. This is a very important factor. The idea of the portfolio is to give a quick *visual* impression of your experience.

I keep descriptive words in the portfolio to a bare minimum. This provides the excuse for me to explain a project—if the client expresses interest. When there is no interest in a particular project, I move on quickly. I use nonverbal cues to gauge the level of appeal. Even though most clients are unaware of it, their eyes open wider when they are interested. Often, clients lean forward for a closer look when something appears relevant to their own needs. At times, clients have grabbed the portfolio out of my hands when they come across "their exact problem!"

A few words about the portfolio format: First, I use color in as many of the sketches as I can. Second, I include large photos depicting me with the equipment or clients. Third, the sketches and materials are neatly arranged and professional looking. Fourth, I usually omit the name of the client for whom I performed the project. This protects (to some degree) my clients' right to privacy.[4]

As a matter of practice, I place my company logo at the bottom of each portfolio page. There are two reasons for doing so. First, if the client wants a photocopy of a particular sheet, I already have my company name on it.

 ## Worth Its Weight in Gold

Many engineering managers are *visual thinkers*. An excess of words, either spoken or written, can turn them off to your sales efforts. With your secret weapon, you get a chance to show them *more* of what you can do. If you tried to do this verbally, clients will often run out of time before you have said enough to convince them.

---

[4] If you have worked on classified or company proprietary projects, you may not be able to include them in your portfolio, even if you delete references to the company and project names.

This is a sound advertising policy that identifies the work as mine. When clients come across it six months later, they won't wonder who authored it. Second, I get double duty from my portfolio by enclosing photocopies of relevant sheets with my proposals. They add instant visual appeal to proposals that contain mostly text and tables.

Examples 1 through 8 give a concrete image of the portfolio format. I invite you to create your own portfolio, one that displays *your* talents and experience in the best possible light. With readily available computer graphics programs and high-resolution printers, making a professional-quality portfolio is a snap.

EXAMPLE 1.

COOLING SYSTEM DESIGN

The Problem:
Design a cooling system for an
airborne infrared surveillance pod.

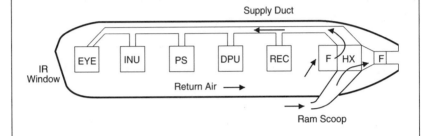

FLOW MODEL

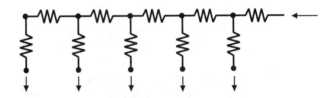

The Solution:
Various options for cooling were evaluated.
A ram air cooling system was chosen as the best overall
performer. Fans and compact heat exchangers were
designed to cool the individual modules.

EXAMPLE 2.

The portfolio concept can be extended many ways. You could bring in *models* of your work or actual hardware, if they are small enough to conveniently pack into a briefcase. Or, using the latest portable computer technology, you could even create an *electronic portfolio*. Generate the pictures with a graphical presentation program. Then create a miniature slide show that

NON-EQUILIBRIUM BOUNDARY LAYER OVER A BLUNTED CONE

The Problem:
Find the electron flux in the wake of a blunted cone as it reenters Earth's atmosphere.

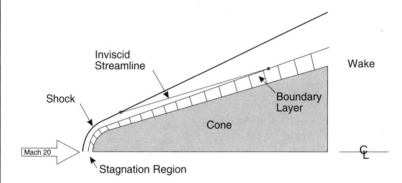

The Solution:
Electrons are produced in the hot stagnation region and carried downstream in the boundary layer. As they move aft, the electrons:
(1) become chemically attached; or
(2) diffuse to the cold wall; or
(3) are carried into the wake.
A chemical non-equilibrium boundary layer program was written to predict the electron flux carried into the wake. Comparison with radar data was excellent.

HK Harvey Kaye & Company

EXAMPLE 3.

highlights your capabilities. Play it back on your *color* laptop at the sales meeting. (I strongly recommend high-visibility screens, such as active matrix color, for this purpose. Low-intensity screens are not sufficiently viewable for presentations, especially when more than one person is watching.) There are four caveats to observe when using computer-based presentations:

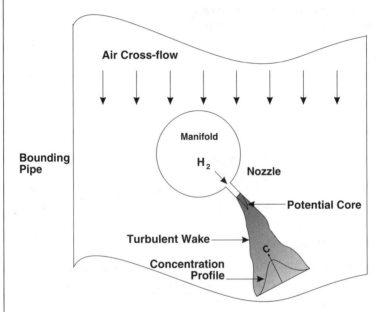

MIXING TEE DESIGN AND TEST

The Problem:
Design a manifold to completely mix a stream
of hydrogen with air, considering pressure drop,
flow, and space limitations.

Air Cross-flow

**Bounding
Pipe**

Manifold

H_2

Nozzle

Potential Core

Turbulent Wake

**Concentration
Profile**

View along end of manifold. Only one nozzle shown.

The Solution:
Jets of hydrogen from the manifold were introduced at an angle
to the air flow. The resulting cross-flow was modeled, and the
turbulent entrainment and concentration profiles calculated.
Later flow tests confirmed the predicted patterns.

HK  Harvey Kaye & Company

EXAMPLE 4.

1. Be aware that some clients don't like electronic presentations. They
   feel more comfortable with old-fashioned "hard copy." Paper gives
   them more control. They can hold it, examine it closer, and more easily point to details. Watching computer images forces clients to a more
   passive role, which means that some of them will be less excited about
   your pitch.

THERMAL ANALYSIS OF CRT ELECTRON GUN ASSEMBLY

The Problem:

A new CRT electron gun design was giving very low production yield. The designers needed to know if differential thermal expansion within the "stack" was causing Plane 1 to buckle into the cathode.

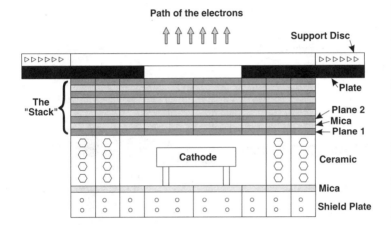

The Solution:

A SINDA thermal model was constructed to assess the temperature distribution. Differential thermal expansions were calculated. Classical buckling theory showed that Plane 1 would not buckle into the cathode.

HK Harvey Kaye & Company

EXAMPLE 5.

2. Use a very fast computer.
3. Make sure your software has the ability to skip backward and forward through the presentation. Don't force the client to wait while your computer presentation churns away.
4. Bring your hard-copy portfolio along in case your equipment fails or the client is not receptive to computer-based presentations.

THERMAL-HYDRAULIC DESIGN OF FFTF TANK

The Problem:
Find a way to dump high-temperature sodium into a storage tank with an initial volume of cooler liquid sodium, so that no part of the tank wall exceeds specified temperature limits.

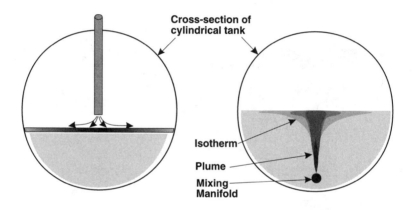

Cross-section of cylindrical tank

Isotherm
Plume
Mixing Manifold

"Before": Hot sodium spreads laterally on top and reaches the tank wall.

"After": A manifold is designed to mix the hot sodium as it rises. The tank wall remains cool.

The Solution:
Design a mixing manifold to distribute the hot sodium coming into the tank. Based on buoyant plume behavior, mixing effectiveness and temperature distribution were calculated to evaluate performance.

HK Harvey Kaye & Company

EXAMPLE 6.

ELASTIC-PLASTIC STRESS ANALYSIS

The Problem:
Determine the adequacy of a PWR spent fuel shipping cask for fire accident conditions.

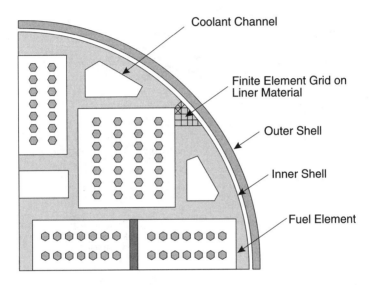

Coolant Channel

Finite Element Grid on Liner Material

Outer Shell

Inner Shell

Fuel Element

One-Quarter Symmetry Section for 2-D Model

The Solution:

A 2-D finite element model of the shipping cask was prepared with ANSYS. The computed temperature distribution was fed directly into the elastic-plastic routines. ANSYS then predicted the peak deformations and stresses in the aluminum liner.

HK Harvey Kaye & Company

EXAMPLE 7.

FINITE ELEMENT HEAT TRANSFER WITH ABLATION CAPABILITY

The Problem:

Add ablation capability to an existing finite element heat transfer code.

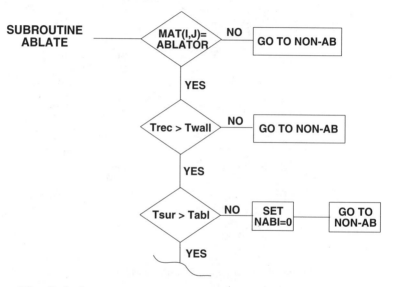

The Solution:
1. Develop ablation algorithm
2. Write FORTRAN code
3. Integrate with existing program logic
4. Debug
5. Verify against proven solutions
6. Write user's manual

HK Harvey Kaye & Company

EXAMPLE 8.

WHAT TO DO AFTER THE MEETING

After each sales meeting, write yourself a set of notes and debrief yourself. Don't trust these important tasks to memory, because sooner or later, you will forget, no matter how good your memory is.

As a minimum, enter new contact names and phone numbers into your database. Do this even if no sale results from the meeting, for contacts eventually migrate to other organizations where a sale may be possible. Note any special data that arise, such as special interests or needs within the client company. Also, file any materials you received such as client reports and drawings that may help at a later date.

If the sales meeting was a bust, jot down *why* it was so — *in ten words or less.* Keep your words very brief, or you probably won't feel motivated to continue this practice over a period of years. Typical jottings might be "Client already committed to another consultant." Or "Need not immediate, but for next year." Or "Fumbled when they asked about object-oriented programming." Or "Couldn't convince them I had sufficient experience in cryogenic cooling." The reasons that fall under your control are vital *feedback* to improve your sales process for future clients.

If the sales meeting was positive in tone, jot down what they liked. Which sales techniques did you handle particularly well? To what points was the client especially receptive? Mention the individuals with whom you developed a strong rapport. Take out your list of action items from the meeting and consider your strategy to handle the next step. Should you be asking for more information, starting work on a proposal, or showing up at the client's door in two weeks? Do you have the necessary information and materials to fashion your next response? If not, then make an action list and get going!

Finally, if you were rejected unfairly or treated shabbily at a sales meeting, meditate for a moment about the larger picture. Recall your passion for your technical specialty and your desire for professional independence. Recall your moments of success, moments when your ability and desire to serve shone through. Recall opening the fat checks you have received as a consultant, checks that prove many other clients regard you as competent and offering valuable service. Next picture that client who tried to deprive you of your dignity. There he is,

in a little canoe, moving up a creek without a paddle. Visualize him in all his foolhardy glory as you wave good-bye. Then release him and move on to other clients who are inclined to treat people with more respect.

GUIDELINES FOR SUCCESSFUL SALES MEETINGS

To summarize, I would like to leave you with the following guidelines for successful sales meetings:

1. Learn to listen! At least three times during a sales meeting, ask yourself, "Am I listening carefully enough?"
2. Be calm. Don't fidget.
3. Do not feel that you have to solve your client's problems while you are sitting in front of her.
4. If you are asked to wait when you first arrive, do not be upset. Bring a small notebook and busy yourself writing in it. If you sense the delay is excessive, try to determine the problem. Many managers unconsciously use delays and interruptions to intimidate subordinates and outsiders. Don't let this happen to you.
5. Do not be too eager to commit your time right on the spot.
6. Try to determine the competitive basis of your bid. Is the client asking many other consultants to bid on the same task, or are you the sole source?
7. Learn to recognize when your brain is being picked without any intent on the client's part to give you a contract.
8. Discuss your past and present projects with enthusiasm. If you have no projects at the moment, give the impression that you are busy regardless, or that you are taking a break from your busy schedule. (Psychology: If you are totally unoccupied, the client's subconscious reaction will be that you are not very good. If he knows you are busy, he will tend to feel that you must be competent and worthy. This general psychology also holds true within organizations—work flows to the people who are most overloaded!)
9. Be prepared for some customers to display negative feelings. Some people regard consultants as pirates.
10. Never bluff! If you don't know a specific fact or reference, either say so or keep quiet. I have seen many consultants put "on the rack" because they said things that they couldn't back up.

11. Think before you talk. Clients are impressed most by the sharpness and relevance of the consultant's questions and her ability to grasp complex situations.

12. Determine who in your client's organization is going to make the decision to approve your contract.

13. Conclude the meeting with an agreement for action. For example, "I shall send you a proposal next week" or "Sorry, I don't think I'm the person to help you in this matter."

14. Before you leave, make sure you have given the client your business card and brochure. Moreover, ask for the client's business card as well.

15. Think positively! Realize that the factors associated with obtaining a particular contract are often out of your control. All you can do is give it your best effort and continue to land a reasonable percentage of the contracts you pursue.

CHAPTER 8

Proposals and Contracts

PROPOSALS AND CONTRACTS: YOUR KEY TO WINNING JOBS

A knowledge of proposal writing and contract negotiation is essential for the consultant. A poorly written proposal or contract can limit the consultant to unfavorable working conditions, cripple his cash flow, or require him to do significant amounts of additional work without compensation. The trick to writing effective proposals is to understand both the customer's technical problem and business practices to the degree that you can anticipate the possible obstacles to successful completion of the work and address the issues *before* a commitment is made.

 <u>Warning</u>

This chapter gives a preliminary view of proposals and contracts for engineering and technical consultants. *The information in this chapter should in no way be construed as legal advice.* If the reader is faced with a particular problem having unclear or undesirable clauses or leaving important issues unaddressed, he or she should seek competent professional advice.

WHAT IS A CONTRACT?

A contract is an agreement between two competent parties, based on mutual promises, to perform specific actions that are neither illegal nor impossible. A contract is more than an agreement. You and I can agree that it's a nice day outside, but that's not a contract. A contract is an agreement with legally enforceable obligations. To be legally enforceable, a contract must contain four elements:

1. Mutual assent
2. Competence of parties
3. Consideration
4. Valid subject matter

"Mutual assent" means that both parties agree to the same set of specific actions and conditions. The consultant and client may go through discussions, negotiations, proposals, and counterproposals to arrive at an agreement whose terms are mutually acceptable.

"Competence of parties" means that both parties must be legally capable of meeting their obligations under the statutes of contract law. For example, contracts with minors, intoxicated individuals, mentally incompetent persons, and enemy aliens are not legally binding in most cases.

"Consideration" is the exchange of service or merchandise on the part of one party for something of value (usually money) on the part of the other. The idea here is *mutuality.* Both parties must give some thing, action, or forbearance (a promise of inaction) that is of value. If only one party gives consideration, it is not a contract, but a unilateral gift.

"Valid subject matter" refers to the legality of the goods and services covered by the contract. You can't make a legally valid contract to rob a bank, to defraud another person, or to suppress criminal evidence. One aspect of "valid subject matter" is much more relevant to consultants. Contracts usually specify which state/national laws and codes are applicable in the event legal reconciliation is necessary. This is sometimes significant for consultants who contract with companies outside of their home state. They may need a P.E. (professional engineer) license in the state for which the contract is valid, even if they already have a P.E. license in their own state. For more details on this, see John Constance's book cited in the Suggested Reading List.

QUOTATION LETTERS

There are two common ways in which individual consultants are requested to propose services. In the first, the client asks the consultant to submit a *proposal* to consider a specific problem. In the second, the problem may not be clearly defined, and it may make more sense for the client to buy the consultant's time. In the latter case, the client usually requests a *quotation letter* to officially transmit the rate to their own purchasing or contracts department.

A typical quotation letter is given in Example 9. If you know the name of the appropriate person in the purchasing or contracts department, you should include it in your letter. In many large companies these letters start the client's internal requisition process leading up to a purchase order or contract.

In your quotation letter, mention how far into the future the rate will remain in effect. A single rate valid over a calendar year is a reasonable approach. Many companies additionally request rates or rate projections valid over the duration of a multiyear contract.

JOHN SMITH ASSOCIATES
2 FORBES ROAD
NEW YORK, NY 21225
212-435-6500

January 5, 1998

ABC Engineering Company, Vacuum Division
1 Fun Drive
Nowhere, CA 94333

Attn: Mr. Charles Doe, Engineering Manager
       Mr. Sam Jones, Contracts Department

Gentlemen:

In response to your request for quotation on heat transfer consulting services on the Delta Vacuum System, my 1998 rate is $100.00 per hour. Any authorized expenses, such as travel for the interface meetings, will be billed at actual cost plus 15 percent G & A fee.

Enclosed is my company brochure and a résumé of my qualifications. Also enclosed is a technical paper I wrote about a vacuum system similar to yours, and sketches of vacuum systems I designed for two other clients.

Thank you for your courteous reception at our meeting yesterday. I look forward to working with you on this project.

Very truly yours,
John Smith

EXAMPLE 9. Quotation letter.

WHAT SHOULD BE INCLUDED IN A PROPOSAL?

A typical consulting proposal contains the following sections:

Introduction

The introduction should persuade the client that you understand the problem and the important contributing contexts. Use language that shows you think "the way the client thinks." Build rapport by stating your interest in helping the client solve the problem. Define the background and discussions leading up to the proposal. State the purpose and goals of your involvement in the client's project.

A proposal is more than a simple statement of work and cost; it is also a marketing instrument. Therefore, make your proposal as persuasive and alluring as possible. Just as in the sales meeting, let your proposals vividly demonstrate the benefits the customer will obtain.

Statement of Work

The statement of work is a clear outline of the steps involved in performing the work. It explicitly describes the services, calculations, tests, computer programs, and reports that you will perform. The statement of work should contain sufficient detail for the client to evaluate the suitability of your approach, but not so much detail that the client can do the work without you.

In most instances, the statement of work should carefully *define* and *limit* the technical scope. For example, if equipment is being designed to operate only under specific conditions, *state* those conditions. If a report is promised, specifically describe its scope; for example, "The report will cover the following six aspects of the project."

The statement of work is also where you specify the frequency and scope of progress reports. For projects that span many months, these intermediate reports help emphasize your continuing presence and can be used to highlight your contributions.

Price and Delivery

This section answers the following questions: How much does it cost? When will it be completed? Who will do the work? What is the basis for the price?

If the project stretches over many months, your price and delivery section should include a schedule and a description of the significant milestones. For complex proposals involving many project phases and workers in addition to you, include a Gantt chart, which shows the start and completion dates of each phase as well as the resources that will be used.

Coming up with a schedule and the number of hours for each task is not trivial. *Estimating* is an art. You need to determine how many hours it will take you to do a project *before* you do it! These uncertainties are compounded because you may be working with a new client under circumstances you may not fully appreciate. Nevertheless, estimating can be made easier by using four simple steps.

1. Break down the project into components.
2. Assign the approximate number of hours it will take you to complete each task, assuming good progress.
3. Add in time for client relations, travel, research, purchasing necessary equipment and materials, and making phone calls.
4. Add the total and multiply by your contingency factor. (I use a factor of 1.5.) Technical consulting projects are rarely as simple as they first appear.

Billing Terms

This section tells when and how frequently you will invoice the client, and how soon he should respond with payment. For projects that last longer than a month, it is common to submit a bill at the end of each month, based on work performed during that month. In cases where there is some risk of not being paid, I suggest invoicing the client every two weeks or asking for partial payment in advance. "Net thirty days" is the standard payment term most consultants use. Some purchasing agents offer terms that include a discount for prompt payment (for example, "Net 30, -2% 10 days"). I always turn them down.

Billing travel time is not always straightforward, because there are few universally accepted policies. If you must travel to visit the client or other facilities on behalf of the client, you should address the matter in your proposal. The issues are:

1. Some clients have rigid rules about expenses such as plane fares, auto rental, hotel accommodations, and meals. For example, certain companies will reimburse plane fares only if they are economy class. Other companies impose limits on auto rental costs. Yet others have per diem limits or "guidelines" for meals. Some companies will not allow you to tack a general and administrative fee (G & A) onto your expenses. In doing so, they place the time and overhead cost of dealing with travel agencies, hotels, and auto rental companies on you. There's not much you can do to counter this situation directly. I recommend trying to recoup your G & A costs by padding the proposal somewhere else.

2. The *time* you spend traveling should be explicitly mentioned as billable time. If you travel cross-country for a monthly client interface meeting on the first Monday of every month, that means you will fly out on a Sunday. Some companies think that "Sundays are free." They even extract this freebie from their own employees. To avoid this problem, explicitly state, "We bill actual travel time, *portal to portal.*" This last phrase means that you will bill the client for

   - The time it takes you to get from your house to the airport
   - The actual flight time inside the plane (but not the time you wait in the airport)
   - The time it takes to get from the airport to your destination, usually the client's facility or your hotel room (if you are flying in the night before your meeting)

   Assuming your hotel is close to the client's site, you would not charge for the time to get from the hotel room to the client's facility.

3. For shorter distances, I drive my own car. My policy is this: If the distance is less than an ordinary commuting distance (about twenty miles each way), I do not charge for my travel time or automobile costs. If the distance exceeds sixty miles, I charge for travel time and automobile mileage costs. In the few instances between these limits (say forty miles), I negotiate. Usually, I ask for travel time but not automobile mileage.

Some consultants impose a minimum charge of half a day (or a full day) for jobs on which they travel, regardless of how little time is spent. Even if the visit lasts only an hour, the client is charged for four (or eight) hours. Other consultants maintain special charge rates for work that involves travel, much as lawyers charge a higher rate when they represent you in court.

Finally, some consultants use travel time on airplanes to work on the client's project. This tends to minimize client objections about charging for travel time. Portable computers have made in-flight work a practical reality. (I must say, though, that I have often seen consultants playing computer golf and arcade games during part of the flight.)

Other Contractual Matters

In this section, address proprietary rights, patent assignments, governing codes and regulations, and outstanding issues that have not been discussed elsewhere. This is also the place for your *disclaimers,* where you state the limit of your responsibility regarding specific issues on the project. If, for example, you are explosion testing equipment, you might want a disclaimer that you are not responsible for refurbishing the equipment after the testing is complete. Or, suppose you are designing a piece of equipment that is intended to be operated in a controlled thermal environment. You might want to disclaim responsibility for the unit's performance if the ambient temperature exceeds specified design limits. Other common disclaimers refer to control of proprietary information and determining who has access and the "need to know."

The *basis* for pricing proposals is perhaps the most difficult issue for the beginning consultant. The generalizations offered on this subject in the next three sections are my own personal observations. Each consultant will find that her own pricing basis depends on her clients' methods of doing business and on the particular nature of her own service.

FIXED-PRICE-BASIS PROPOSALS

If the consulting service involves a well-defined task, the client may want to put a ceiling on the maximum expenditure. The result is a request for a firm, *fixed-price quotation*. In fixed-price proposals, the client only wants to know what it costs to have the job done. He is not concerned about hourly rates or how many hours are needed to accomplish the task. A firm, fixed-price contract is a promise to do the specified scope of work — no excuses allowed! Therefore, the consultant must specify *exactly* what the scope of the work is. If this is not done, the client can come back and ask for "related" work at no expense. These freebies can easily exceed the original scope of the contract, resulting in a very low effective rate of pay. I know of many consultants who have ended up working for five dollars per hour on a fixed-price job. So, exercise great care with the statement of work on fixed-price bids!

Tip

If there is *any* aspect of the job that contains an unspecified interface, a yet-to-be-determined part, or a major variable that is unknown, don't bid on a fixed-price basis! Either bid on the *part* where you know all the variables, or bid the whole project on a time-and-materials basis. Time and time again, I have seen consultants rationalize to themselves, "That one little part *couldn't* be that bad; I'll handle it when I get there" — only to find out that it *could* be that bad. The hallmark of a *development* effort is that the major variables are poorly defined. Such jobs should be bid accordingly, that is, on a time-and-materials basis.

EXAMPLE 10. Fixed-price proposal

JOHN SMITH ASSOCIATES
2 FORBES ROAD
NEW YORK, NY 21225
212-435-6500

July 1, 1998   Proposal 128

ABC Industries
1 Gulf Road
Boston, MA 01153

Attn: Mr. Charles Doe
Subject: Proposal for Thermal Testing of ABC Model VI Corn Popper

Dear Mr. Doe:

We are pleased to submit this proposal for the thermal testing of your Model VI corn popper. As we discussed in our meeting on June 18, 1998, the testing will be done on the three prototype models to be furnished by ABC. These units will be returned to ABC after completion of the tests.

As a result of our testing, you will be able to pass your Underwriters Lab certification. Our test data will also provide useful design feedback for developing the forthcoming Model VII. In response to your concern for fast results, we are offering a two-week turnaround on this project, which is significantly faster than that offered by other vendors.

Scope of Work

For each of the three prototypes, the following tests will be performed:

1. Heat-up transient with three ounces of popping oil; measure oil temperature versus time
2. Heat-up transient with no oil; measure plate center temperature versus time
3. Maximum plate temperature nonuniformity versus time with three ounces of popping oil; measure temperature differences and locations

The results will be collected on our data recorder, converted into Excel spreadsheet format, and plotted on charts generated by Freelance Graphics. A test report will be supplied to ABC, giving test configuration details, relevant test conditions, results, and conclusions.

Price and Delivery

The proposed effort will take two weeks to complete after receipt of your purchase order and the prototype units. The work will be performed by John Smith with the lab staff assisting as required.

The cost of this effort is $4,000.00 on a firm fixed-price basis. If additional testing beyond the scope of work is required later, its cost can be negotiated at such time.

(continued)

Terms are net 30 days. Cost and delivery quoted in this proposal are firm for 90 days, after which John Smith Associates reserves the right to revise them.

Attached is a copy of our Standard Fixed-Price Terms.

If you need more information on any aspect of this proposal, please do not hesitate to call.

Very truly yours,
John Smith

Enclosures

EXAMPLE 10. (continued)

Another aspect of fixed-price work is that the emphasis is on getting the job done, and done *correctly*. If the consultant makes an obvious and significant error on a fixed-price job, he may be legally or ethically bound to correct his work at no additional cost to the client.

A good rule of thumb for quoting fixed-price work is that the fixed price should be about twice your best estimate of the time and materials involved.[1] Much time will be lost in talking with the customer, looking up references, writing letters, and in the start-up activities that are inevitable on any new project.

Fixed-price proposals should be made valid for a specified number of months after the date of quotation. Since clients often change the design and system parameters during a project, structure the wording so that a follow-up contract can be written to accommodate any new developments resulting from the original contract. Example 10 shows a typical fixed-price proposal.

The art of writing fixed-price proposals is similar to devising an à la carte dinner menu. By carefully defining and limiting the entrees you promise, you can sell the salads, desserts, and after-dinner drinks as well.

Some consultants make an extremely good living using only fixed-price bidding. These people all have the advantage of offering a service in which they are far more proficient than the customer would ever guess, and where they have good control of their testing, computer, and equipment costs.

THE "MENU" PROPOSAL

The menu proposal is a variation of the fixed-price proposal that offers the client a number of possibilities to choose from. It is based on the old Sears

---

[1] Generally, you should not tell the client the estimated number of hours it would take to finish a fixed-price job. The customer is concerned with the bottom-line cost, not with the amount of effort it takes to get there.

Roebuck strategy of offering "good," "better," and "best" models addressed to different market segments and pocketbooks. Preparing a proposal of this type requires more work but gives you a greater chance of landing a contract because clients can choose the offering that most closely matches their needs. The key to success with this type of proposal is to

- sit down with the client and ask what details they visualize for each option;
- include only those options that you are capable of handling;
- do not price the *details* of each option, only its overall price. This counteracts efforts by the client to nitpick your price structure. ("Gee, on option A you're charging 2K for a user's manual, whereas on option B, you want 5K. Is there really that much difference between the two?")

The menu proposal in Example 11 is greatly abbreviated and is intended to illustrate the general format and layout only.

TIME-AND-MATERIALS-BASIS CONTRACTS

In situations where the scope of the technical problem is not well defined, the consultant may be asked to propose on a time-and-materials (T & M) basis. The client is simply buying the *time* the consultant has expended and the *materials* he has consumed. See Example 12.

The T & M contract is made on a "best efforts" basis. An *estimate* of the time to complete each item in the scope of work is prepared. There is no *guarantee* of completing the work in the estimated hours, and no promise to "make it right" if the consultant (or the client) has erred in data, methods, or assumptions. Many large consulting companies are notorious for bidding low on T & M contracts to get their foot in the client's door. They then overrun the original estimates by factors of five and ten. This is not a recommended practice for the individual consultant. Realistic estimates will help to maintain your credibility over the long haul.

Most companies issue purchase orders stating that the T & M contract can be halted on short notice. Thus, obtaining a T & M contract for fifty thousand dollars is not a guarantee that your billings will reach that level. The project may be canceled, or other conditions may necessitate terminating your efforts prematurely. If you must rent or purchase equipment, or otherwise commit your money, in order to perform a T & M contract, it is a good idea to negotiate a *termination clause* with your client so you won't get left holding the bag on obligations you have incurred.

EXAMPLE 11. Menu proposal

JOHN SMITH ASSOCIATES
2 FORBES ROAD
NEW YORK, NY 21225
212-435-6500

July 1, 1998   Proposal 144

Super Scintillating Software, Inc.
3 Gates Road
Macedon, CA 06543

Attn: Mr. Peter Strass
Subject: Proposal for Excalibur Shop Floor Control Software

Dear Peter:

We are pleased to submit this proposal for the Excalibur Shop Floor Control Software. As we discussed in our meeting on June 12, 1998, the demo version is intended primarily as a sales tool to quantify the demand for the system *before* a large capital investment is made. The demo will be an animated slide show created with Microsoft Powerpoint and will use dummy screen shots to walk viewers through and demonstrate the functionality and features of the system. The shareware system is a bare-bones functional system without user manual, import-export capability, and report-generation functionality. It can be offered through Internet shareware channels as an inexpensive introductory version. The commercial version is the full-blown product with all features included. It will satisfy the needs of the most demanding industrial user, providing the basis for a complete shop floor control system with documentation and extended data-sharing capabilities.

Scope of Work

The scope of work is summarized in the following table and in Attachment I. Notes:

1. Beta testing includes bug fixing up to the release date. After that point, bug fixes and recompiles are not included in the price and must be obtained separately from John Smith Assoc.
2. User manual, Master 3.5 inch floppy disks, and CD will be furnished reproduction-ready. The costs of reproduction are *not* included in this proposal.
3. Details of the beta-testing program follow our standard procedures and are given in Attachment II.
4. Commercial packaging includes design of the box, its artwork, and advertising copy in service-bureau-ready condition as outlined in Attachment III. It does *not* include reproduction costs.

Item	Demo	Shareware Version	Commercial Version
Walk-through demo	Y		
Two example cases	Y	Y	Y
Specification development		Y	Y
Graphical interface		Y	Y
Expert engine integration		Y	Y
Database import/export			Y
Report-generation features			Y
Registration & software locking		Y	Y
On-line help system		Y	Y
Beta-testing program		Y	Y
User's manual			Y
Commercial packaging			Y
Master distribution disks		Y	Y
Master CD for commercial version			Y
Cost	$15,000	$75,000	$175,000
Months required to deliver	2	4	6

Price and Delivery

The proposed work will be completed in the delivery time tabulated above, measured from the time of receipt of your purchase order. The work will be performed by John Smith with the programming staff assisting as required. The "deliverable" will consist of complete source code for the software and all supporting documents. If applicable to the option selected, the deliverable will also include beta test results and distribution media master copies.

The costs given on row 15 of the above table are quoted on a firm fixed-price basis. If additional effort beyond the scope of work is required later, its cost can be negotiated at such time.

Terms are net 30 days. Cost and delivery quoted in this proposal are firm for 90 days, after which John Smith Associates reserves the right to revise them. Attached is a copy of our Standard Fixed-Price Terms.

If you need more information on any aspect of this proposal, please do not hesitate to call.

Sincerely,
John Smith

Enclosures

EXAMPLE 11. (continued)

EXAMPLE 12. Time-and-materials proposal.

JOHN SMITH ASSOCIATES
2 FORBES ROAD
NEW YORK, NY 21225
212-435-6500

July 1, 1998   Proposal 129

Telescope Corporation
1 Atlantic Street
Boston, MA 01150

Attn: Mr. Henry Vincent
Subject: Proposal for Thermal Analysis of the WHM Telescope

Dear Mr. Vincent:

This letter proposal describes our approach to the thermal performance analysis of the WHM telescope.

The WHM telescope is a cryogenically cooled infrared telescope intended to operate in low Earth orbit. It must accommodate a number of thermal loads including solar, albedo, and internal heat generation while maintaining the optical surfaces at specified cryogenic temperatures. To determine its performance analytically, we propose the following scope of work.

Scope of Work

1. Create a computer thermal model of the telescope using the Aardvark computer code. The Aardvark code is an industry-standard tool using finite elements to model the thermal behavior of complex systems. The model of the WHM telescope will utilize temperature-dependent thermal conductivities, cryogenic enthalpy data from the Sigmon tables, and transient response capabilities. The telescope will be modeled with about fifty elements. All dimensions and assembly data will be based on Drawing AC-581 in your Request for Proposal (RFP). Representation of the following will be included in the model: mirrors M1–M5, scan motor, front baffle, front and rear decks, tube, flange, and support structure.

2. Determine first-order transient thermal loading, including: solar, albedo, Earth radiation, internal heat generation, and heat leakage from vacuum vessel. These loadings will be based on the telescope configuration and operating scenario described on pages 121–124 of your RFP.

3. Run the computer model for three different tube materials: 6061 aluminum, 1100 aluminum, and copper.

4. Prepare a complete analysis report, giving conditions, computer model parameters, results, and recommendations.

Cost and Delivery

The work described in this proposal is quoted on a time-and-materials basis at an estimated cost of $24,000.00. This cost will not be exceeded without prior approval from Telescope Corporation. Services will be billed at the following rates:

- Senior Engineer $100.00/hr
- Engineer..$75.00/hr
- Engineering Assistant $40.00/hr

Professional résumés of John Smith, Senior Engineer, Cyrus Quick, Engineer, and Marvin Kulitch, Engineering Assistant, are included as an appendix to this proposal.

Computer costs will be billed at cost plus 15 percent general and administrative fee. The work will be completed 90 days after receipt of your purchase order and required drawings. Billings will be submitted at the end of each month, based on work performed during that month. Terms are net 30 days.

Proprietary Matters

John Smith Associates understands the proprietary nature of the WHM telescope and is willing to give reasonable and proper assurance that information regarding it will be safeguarded.

Should any further details or explanation be required, please contact me or Cyrus Quick at our office. We hope to hear from you soon.

Very truly yours,
John Smith

EXAMPLE 12. (continued)

MEET THE PROS UP CLOSE AND PERSONAL

Ted Husted is the owner of Husted.Com, an Internet consulting and network support services company. Ted started in the computer programming field about ten years ago, where he gradually developed an interest in electronic book publishing. He created a number of  hypertext programs to serve as containers for electronic books and published a few of these programs as shareware products. When the Internet became popular, Ted had a head start in HTML (HyperText Markup Language) because it used much of the same technology he had already developed for his own programs. After a one-year transition period at his last employer, Ted went into full-time consulting in 1994.

Because the Internet was brand-new, Ted's market was "moving in fifty directions at once." Ted did some experiments to discover what types of clients and contracts worked best for him. The results were very interesting: Ted found that he could support ten small clients (with five machines each) much more easily than one large client with fifty machines. That is, the scaling was not linear for this type of consulting. Therefore, Ted targets only smaller companies for his Web-development consulting. When and if the client grows beyond a certain size, Ted hands off the support to a larger consulting organization better suited to the higher load.

Ted uses fixed-price quotes a lot in his work. He explains, "Many customers don't understand consulting rates and don't understand that two or three hours of professional education go into every hour of expert advice or support. It's also difficult for many clients to approve a project without firm pricing, which conflicts with the industry-standard hourly rate model. Lately, I've switched to a fixed-price approach and no longer publish an hourly rate."

How does he like consulting? Ted says, "Consulting is a wonderful life, but I've had to learn to live with a six-week work 'horizon.' At any given instant, I know what I'll be doing only for the next six weeks, but after that...New work keeps coming in and pushing the horizon back, but it's always there."

Unlike the fixed-price contract, the T & M contract is easily extended. The client's purchasing department merely adds dollars to the Not-To-Exceed (N-T-E) amount and issues a *change order*. The consultant receives a new purchase order based on the old purchase order number, but identified as Revision 1 or Change Order 1.

I use the T & M contract almost exclusively in my own practice and find it to be a satisfactory arrangement. It has the flexibility to accommodate a number of different situations regarding follow-up work, and it has fewer liabilities in the event of unforeseen difficulties in performing the work.

WHAT IS THE DIFFERENCE BETWEEN A PROPOSAL AND A CONTRACT?

In general, proposals and rate quotation letters are *not* contracts. Although the client may specify that parts of the proposal are to be considered *parts* of the contract, the contract, in most situations, is the paperwork that the client sends back in response to the proposal. In most cases, this is a purchase order or a document prepared by the client's contracts department, called a *consulting agreement*. The contract becomes binding when the consultant signs the "acceptance" copy of the purchase order or consulting agreement.

Make sure that you agree with the contract terms *before* you sign an acceptance! For example, you may have proposed that billings be on a monthly basis, and you find that the client's purchase order specifies payment at the end of the project. If this is not acceptable, you have every right to negotiate this point and have the client draw up a revised document.

Many companies use the *purchase order* as the contract. The front of the order has the statement of work, rate of pay, N-T-E dollar limits, etc., and the back side has the company's general terms in fine print. The company in this case will include an "acceptance" copy for you to sign and return, after which time the contract is binding. To be valid, the client's purchase order must be signed by a person with *contractual authority*. This power is usually given to the head of the purchasing department, the manager of contracts, and most company officers. If the contractual authority of the person signing is not apparent, call and verify it. Of course, an *unsigned* purchase order, like an unsigned check, is worthless.

Some large companies issue both purchase orders and consulting agreements. In most of these cases, the consulting agreement is the official contract because it is signed by both parties. The purchase order from the purchasing department gives you a P.O. number and other instructions for billing that are usually not stated in the consulting agreement. Examples 13 and 14 show typical formats for consulting agreements and purchase orders.

EXAMPLE 13. Typical consulting agreement.

CONSULTING AGREEMENT

This Agreement is made between ABC, Inc., a Massachusetts corporation (hereinafter referred to as "ABC") and John Smith (hereinafter referred to as "CONSULTANT"). ABC agrees to contract for the services of the CONSULTANT, and the CONSULTANT agrees to provide services under the terms and conditions in this agreement.

I. Statement of Work
The CONSULTANT shall provide consulting services on behalf of ABC in the area of thermal software development (see attached description) and any related matters ABC may request during the period of performance.

II. Payment for Services
In full consideration of the consulting services hereunder, ABC agrees to pay CONSULTANT at a rate of $100 per hour, plus reasonable expenses incurred at the request of ABC.

A monthly invoice describing services rendered and expenses incurred shall be submitted to ABC at the end of each month in which the services are rendered.

III. Period of Performance
CONSULTANT shall be available for a maximum of 200 hours (25 days) for the period of performance beginning Jan. 1, 1998, and ending Sept. 30, 1998. This period of performance shall not be extended without written authorization by ABC.

IV. Not To Exceed (N-T-E) Limit
Total payment under this contract shall not exceed $20,000, unless authorized in writing by ABC.

V. Independent Contractor
It is understood and agreed that: CONSULTANT is an independent contractor in the performance of this Agreement, consultant is not an agent or employee of ABC, and CONSULTANT is not authorized to act on behalf of ABC.

CONSULTANT shall assume full responsibility for payment of all federal, state, and local taxes, and/or special levies required under unemployment insurance, social security, income tax, and/or other laws, with respect to performance of the consultant's obligations under this Agreement.

VI. Right To Act as Consultant
CONSULTANT warrants to ABC that he is not subject to any obligations, contracts, or restrictions that would prevent him from entering into or carrying out the provisions of this Agreement.

VII. Termination
This agreement may be terminated by either party at any time by giving written notice of such termination to the other party. Upon receipt of such written notice

---

(continued)

by either party, no further charges will be made under this Agreement. Termination shall not affect the CONSULTANT's obligations under articles IX and X.

VIII. Hold Harmless

CONSULTANT shall indemnify and hold ABC harmless from any and all suits, claims, actions, damages, or losses whatever, resulting from any act or omission of the CONSULTANT, its employees, agents, and subcontractors in its performance hereunder.

IX. Confidentiality

CONSULTANT acknowledges that information about the research, design, development, marketing, and manufacture of ABC's products, including findings, reports, and improvements made or conceived by the CONSULTANT under this Agreement, is confidential and of great value to ABC. Accordingly, CONSULTANT agrees not to disclose any such confidential information to any person not authorized by ABC to receive it. Upon completion of the work, CONSULTANT shall deliver to ABC all documents, drawings, specifications, and similar materials that were furnished by ABC to CONSULTANT or that were prepared by CONSULTANT in performance of services hereunder.

X. Discoveries, Inventions, and Copyrights

CONSULTANT will promptly disclose to ABC all inventions, improvements, designs, and ideas made or conceived by CONSULTANT in the course of CONSULTANT's services under this Agreement. CONSULTANT assigns to ABC all right and title to such inventions, copyrights, and developments, and agrees to execute any and all such documents, including patent assignments, as ABC deems necessary to secure to it all right, title, and interest.

XI. Amendments

This Agreement may be amended only by a written document, signed by both ABC and CONSULTANT.

XII. Assignment

CONSULTANT may not assign this Agreement or any right hereunder. Any such attempted assignment shall be void.

XIII. Governing Law

This Agreement shall be governed by the laws of the Commonwealth of Massachusetts.

CONSULTANT	ABC
by_____	by_____
Name_____	Name_____
Title_____	Title_____
Date_____	Date_____

EXAMPLE 13. (continued)

UNIVERSAL INC. 4 TECH LANE WALTHAM, MA 01522 311-654-9807	PURCHASE ORDER NO. _X4321_ REV.00 PLEASE NOTE: THIS ORDER NUMBER MUST APPEAR ON YOUR INVOICE AND ALL PACKAGES
VENDOR John Smith Associates 2 Forbes Road New York, NY 21225	MAIL INVOICES TO: ACCOUNTS PAYABLE DEP'T. UNIVERSAL INC. 4 TECH LANE WALTHAM, MA 01522

ORDER DATE 09-18-92	VENDOR NO. 34778	TERMS Net 30	☐ TAXABLE ☒ NOT TAXABLE

QUANTITY	DESCRIPTION	PRICE
1	Consulting services of John Smith on thermal analysis of Iddy-Biddy control unit, project 176B6 100 hours @ $90.00 per hour Total cost, including expenses, not to exceed $9,500.00	9,500.00
	TOTAL AMOUNT OF ORDER	9,500.00

THIS PURCHASE ORDER IS SUBJECT TO THE TERMS AND CONDITIONS ON THE FACE AND REVERSE SIDE HEREOF

UNIVERSAL INC.
by *Don Jonson*
Don Jonson
Purchasing Agent

EXAMPLE 14. Purchase order.

HOW TO RESPOND TO A REQUEST FOR PROPOSAL

When many companies want to buy equipment or services, they send a *Request for Proposal* (RFP) to qualified vendors on their vendor list. The RFP is, in essence, a specification that tells vendors exactly what the company wants to purchase. The level of detail varies with the item. For a simple PC computer system, for example, an RFP may contain sections on component selection, acceptable fabrication procedures, performance, quality, testing procedures, required drawings and documentation, applicable codes and standards, maintenance, shipping container design, terms for delivery, and vendor credentials.

Similarly, consulting work is sometimes offered through the RFP. Large companies and government agencies know their own needs better than most outsiders. Rather than ask twenty consultants to come in for lengthy sales meetings, these companies prepare a document, the RFP, that defines the problem and tells vendors exactly how to propose on the solution. This approach is generally more cost-effective.

The proposal you write in response to an RFP is just like any other proposal you might write, except that it is structured according to the customer's specific format. This way, the customer is assured that the incoming proposals address all significant issues. There is nothing more frustrating to a large company than receiving proposals, some of which neglect important issue A and others of which neglect important issue B. To avoid the situation of comparing apples and oranges, the RFP specifically mandates the critical items to be addressed.

In responding to an RFP, you will be asked to deliver a written instrument — the proposal — to the company. Therefore, your success in winning these kinds of jobs depends on your writing and your ability to organize a technical approach to complex problems.

Depending on the complexity of the job, you may have to write a

- Synopsis of the client's problem to demonstrate that you understand it
- Technical section showing how you will do the work in sufficient detail so that the client can evaluate it
- Plan showing schedules, staffing, required equipment, labor and cost estimates, proprietary information handling procedures, and report formats
- Company business description (see Chapter 14) that explains your consulting history, credentials, facilities, and bank references

In sum, your proposal may require from five to one hundred pages of material. Once you have completed one such proposal, much of this will become standard boilerplate that can be recycled in other proposals. In responding to an RFP, keep the following in mind:

- Make sure you have addressed *all* relevant items in your proposal. Leaving out trivial items can sometimes automatically disqualify you.
- If the RFP says that proposals must be in by a certain time on a certain date, they mean it. Late proposals are almost always disqualified.
- Contact the company by phone to make sure you have the latest version. Sometimes, after the original RFP is circulated, some bidders discover inconsistencies and press the company to issue an update. If the company neglects to send you the update, your proposal effort may be wasted.
- Market the contact person handling the RFP. Get as much information as you can about going prices, hot issues, and special concerns. Expect resistance.
- Remember that, *by design,* the RFP process means you have competition. Discover what they are doing and make sure you have a competitive edge.
- For the same reason, expect other competitors to be interested in what *you* are offering. Don't give away the "company jewels" in your proposal! The way the competitive bidding process is structured, critical information in your proposal will eventually find its way into your competitors' hands.
- In some cases, responding to an RFP may not be worth it. If your chances of receiving the contract are low, or if the required proposal is very large, you may be better off using the time for other marketing efforts.
- Price is not always the determining factor in selecting the winning bid. In some cases, the company may have no idea of how to proceed with the technical problem. Your unique technical approach or credentials may then be the most critical factor.

KNOW THE TRUE VALUE OF YOUR EXPERTISE!

So far, I have described the three most common ways of pricing your work. Other methods of pricing are also used in technical consulting, but you probably won't need them to start. That is, all except one. And that deals with the situation where the basis of the price is the *perceived value* to the client. In some cases, your expertise has value to the customer that is out of all proportion to the time it takes you to perform the service.

You may know the story about the consultant who was asked to fix a complicated steam-driven machine. This machine was a critical factor in the plant's financial operation; every hour lost would cost the company thousands of dollars. The consultant, who knew this machine better than anyone else, was invited to the plant. He poked around for half an hour. Then, without much ado, he took out a small hammer and hit one of the steam pipes in a particular spot. The machine suddenly came to life again!

The consultant sent in his bill. When the client saw it was for $1,000, he asked the consultant for a price breakdown. After all, the consultant had spent only thirty minutes on the job.

The consultant responded with the following:

- Hitting steam pipe at location A ....$ 5.
- Knowing where to hit steam pipe ....$ 995.

The above situation is not likely to happen to you. However, in some instances, your knowledge will be a hot commodity whose value is not directly related to an hourly charge rate. Recognizing these instances is *extremely* crucial to your success.

When I started out in 1976, I had a friend who was also starting his own consulting business. He had a prospective client with an urgent problem. But my friend couldn't figure out how the problem could be solved. He wrote a proposal to his client using technical arm waving and irrelevant generalizations. The client was not convinced that my friend really knew what he was talking about. The problem involved my exact specialty, so my friend asked me to look at the problem and write a convincing proposal *for* him. When the job came in, he would give me the work to do at an attractive rate. Because we were "friends," everything was verbal and based on mutual trust.

I spent a week figuring out the client's problem, and wrote a comprehensive and convincing technical proposal. It explained why a certain approach was needed, and contained enough technical "meat" to convince the severest of skeptics. My friend got the contract for $300,000. And then, he proceeded to hire a young Ph.D. to do the technical work outlined in my proposal. I not only didn't get to do the work at an attractive rate, I didn't get compensation for writing the proposal!

Well, I was naive then... What would I do if I were in that situation today? First, I would realize that *very specialized knowledge and a hot customer add up to business opportunity and profit potential.*

I would explain to any "friend" or client for whom I wrote a proposal (in which my in-depth knowledge of the subject made me unique) that simply billing the $5 to hit the steam pipe at location A is not adequate reward. I would arrange a *written* understanding for

- hourly compensation for writing the proposal itself;
- a commission of 2 to 10 percent of the total billing to his client, in the event he is awarded the contract.

The commission for writing a proposal depends on how critically the contract hinges on your efforts. If your part of the overall proposal is the essential element in winning the job, then your cut should be closer to the 10 percent figure. If you are a major contributor in a large proposal that has other major contributors, your fee should be proportionally lower.

LET YOUR PROPOSALS SHOW YOUR STRENGTH

In writing proposals, *use* your special knowledge and experience to your advantage. If your client is floundering with a problem or if he can't define his needs clearly, he may be misrepresenting the problem to *all* the consultants asked to bid on the job. You have a clear advantage if you can explain how the job should be defined and performed (in a tactful way, of course).

For example, I was once asked to propose a thermal design for one of the country's largest electronics manufacturers. As I was writing the proposal, my experience with similar projects made me realize that two pieces of the puzzle were missing. Without these pieces, there was no satisfactory solution to the problem. I went back to the client and in a short presentation explained what the missing information was, why it was needed, and how I could go about creating a "mini-specification" for the data to be supplied. I showed how I would use the needed data in my approach and how we could mutually develop it.

I won the contract, even though others were slated to get it. Later the project manager told me, "Your proposal was the only one that tied all the factors together. You were the only one who seemed to *know* what you were saying."

PROTECTING YOUR OWN PROPERTY

Suppose you have developed a special computer program at your own expense and decide to use it for a customer's application. The customer may hire you for a week to calculate input data for the program. But the program took you six months to develop. How are you going to be fairly compensated for its use?

First, you could *sell* your program to the client, describing it as a material expense in your proposal. In this case, you supply the program, source code, and program documentation in a package. Whatever they do with the program after that is their business, since they own it. You might additionally propose support under a labor category to maintain and/or operate the program for them.

But if you don't want to sell the program to them, you could propose a program licensing fee that compensates you for your effort in developing the program. This charge should be proportional to the development effort and the uniqueness of the program on the market. The sticky part of such arrangements is that the client frequently wants a copy of the source code "just so we can verify that it works for standard in-house cases" or "because we have in-house rules that say we must verify all software."

What I recommend in this case is that you prepare a "program description document" that shows, in terms neither too general nor too specific, the program's capabilities, the methods it uses, and its performance on industry "benchmark" cases to demonstrate its validity. This will allow you to charge a significantly lower price than an outright "sell," and the customer still gets what he wants. If you do a substantial amount of this work, consult recently published books on computer program copyright law and marketing procedures.

COMMON SENSE IN LEGAL MATTERS CAN SAVE YOU HEADACHES

Common sense in legal matters is a great asset to the consultant in writing proposals and contracts. It is worth the effort to ask sufficient questions to assess the overall context within which the consulting work will take place. Listen carefully to discover any sensitive aspects of the project.

It would be difficult to enumerate all the special circumstances that "red flag" potential legal pitfalls. Nevertheless, as a minimum, if the work involves any of the following, the contract should address the issues and pass your careful scrutiny:

- Product liability
- Public endorsement of a product's suitability
- Utilization of union labor
- Providing of expert witness for court cases
- Guarantees of results
- Patent fights
- Proprietary or classified information
- Early termination of work

- Penalties for nonperformance
- Use of codes and standards
- Special permits and licenses
- Environmental hazards
- Compliance with handicapped access
- Unusual kinds of insurance you must carry
- OSHA regulations
- Responsibility for consequential damages to client's equipment

All of this must seem totally intimidating to you! Do not lose heart, dear reader. Contract law is so broad and complex that even an expert may not be able to authoritatively advise you on all of the above special situations. The best you can do under the circumstances is to learn the basics and become aware of what you *don't* know. Never be reluctant to seek out advice. In cases where contractual liabilities are either unclear or disadvantageous, it may be best to turn down an otherwise desirable contract.

As a word of encouragement, it should be mentioned that the vast majority of consultants find that contractual and legal matters occupy a minuscule fraction of their attention. In the everyday performance of proposal writing and contract negotiation, an elementary knowledge of contracts is sufficient.

WHEN TO DECLINE WORK

As a beginning consultant, pay extra attention to the contractual and legal aspects particular to your practice. Make sure that your proposal covers them. For example, if you test equipment for your customers as a regular part

Tip

You can often sidestep potential legal pitfalls by making careful disclaimers in your proposals. A tactfully worded disclaimer can save you from getting tied up in a costly court case in the event of a problem or a contract violation. However, that's not the end of the story. You must also make sure that the client *accepts* your disclaimer. That is, *before* you sign the acceptance copy of the contract or purchase order, it must contain your disclaimer *in writing!*

of your business, you may want to include a "standard terms" sheet in your proposals. These standard terms define the limits of your liability for damage occurring to the customer's equipment during the testing and outline how refurbishment and retesting would be costed and scheduled. If the client rejects your standard terms and you can't arrive at a mutually acceptable compromise, you may want to decline bidding on the job.

As another example, suppose you are negotiating a contract to write a computer program for a client. The issue of *acceptability* comes up. The client wants you to guarantee that your program will run on *all* IBM-compatible personal computers. This poses a potential problem, for if you agree to these terms, you might have to modify your program for each and every brand of "compatible" PC that behaves differently from the system you used to develop the program. There are dozens of manufacturers who produce IBM compatibles that deviate from the norm. Small differences in hardware, video standards, memory management schemes, and operating systems all contribute to the likelihood that your program may not run on these deviant systems. You might want to counter with a more limited guarantee: "Will run on an Intel Pentium system with 16 megs of RAM, SVGA graphics, and Windows 98 operating system." If you cannot come to a satisfactory middle ground, turn down the work.

As a final example, suppose you are an expert inventor with many patents to your credit. If you are asked to propose on consulting work that utilizes your genius for invention, hourly compensation may not be sufficient. Counter by specifically stating how patent rights and royalty issues will be handled. Unless you *make* it an issue, the client will assume that their standard terms are adequate. If you do not take the initiative in the proposal stage, the issue will be harder to negotiate when the contract appears. I know many consultants who have turned down this kind of work because they could not come to satisfactory patent arrangements.

With marketing such a formidable task, you would think that you should say yes to every project that comes your way. Yet, there are times when saying no makes more sense:

1. When the project is clearly outside of your expertise and there is not enough time to learn the material adequately.

2. When you are already overloaded and a client asks you for the same time slot. (In this case, though, I would first press strongly to modify the time slot so that I could still meet the client's need. I would even juggle other projects under way if the project represented a significant opportunity.)

3. When the client's expectations are unrealistic. Suppose you think it is an eighty-hour job and the client insists it is a two-hour job. The gulf is so wide that negotiation may not bring you to a satisfactory common ground. In such cases, it is better to say no than to work at one-fortieth your rate. Similarly, if your rate is not acceptable to the client and they want you at one-third of your normal rate, it's time to apply some sales strategy: Show how you offer them more value than the competition. If this doesn't work, politely say no.

4. When a new client asks you to drop everything and help fix their emergency panic. If you have worked for the client before and you know they pay their bills, I would be more inclined to say yes.

5. The project calls for unachievable results or is so poorly defined that you might not be able to give them their money's worth.

6. The project places you in a political bind. If you are doing steady work for a company, don't let their direct competitors sweet-talk you into a small contract. You will appear disloyal to the first company, and they will dump you at their earliest convenience.

7. The project places you at risk.

8. The client is not credible. Some companies are well known for not paying consultants and playing other dirty tricks. Steer clear!

9. The project is not worth the bother. If you have to travel (at your own expense) an hour each way to meet with a client for just one hour's consulting, your effective overhead rate skyrockets. In addition to the travel, you will have to call to confirm the meeting (ten minutes plus phone charges), pack your briefcase with your promotional materials (fifteen minutes), and prepare an invoice (fifteen minutes). Thus, for one hour's billable work, you have put in three hours and forty minutes of effort. To avoid this situation, many consultants have a *minimum* charge of four or eight hours, even if the actual consultation lasts only one hour.

10. You can say no in the middle of a project if you feel the client is severely compromising your position, efforts, or professional integrity, or if the client has misrepresented a significant issue that bears directly on your ability to contribute effectively to the project. This doesn't happen very often — perhaps one project in twenty. Before you say no in this instance, consider your contractual liabilities and the possible repercussions on your reputation.

11. When the project leads you in the wrong direction and you have other viable choices that take you closer to your goals.

WHAT TO LOOK FOR IN VERBAL GO-AHEADS

When you are called into an emergency or rush situation, there is often not sufficient time for the purchase order or consulting agreement to be written. If your client is a large and reputable firm, it is useful to remember that verbal go-aheads, followed by written confirmation, are sometimes used. Generally, the only people authorized to give verbal go-aheads are company officers, the manager of purchasing, and the manager of contracts. It is a simple matter to ask if the person telling you to "go ahead" is authorized to do so. In any case, you will still need the following information verbally:

1. The *number* of the purchase order
2. Verification of the billing rate
3. An N-T-E contract amount or the number of days you are authorized to work on the project
4. An explicit agreement that the paperwork will follow shortly
5. The name of the purchasing agent or authorizer

It is understood that all unspecified conditions in such verbal transactions will follow the client's normal procurement procedures. There is some chance that the standard terms may not be to your taste, so if you have special concerns above and beyond the basic verbal order, it is wise to mention them explicitly before you agree to start work.

I know many consultants who have been burned with invalid verbal contracts. The most common problems are disagreements over the amount of time authorized and "authorizations" made by client personnel who did not have contractual authority. However, the chances of such negative experiences are greatly minimized if the above suggestions are followed.

CHAPTER 9

Billing Rates

WHY MONEY IS A SENSITIVE ISSUE

Have you ever observed that when the talk turns to billing rates, the adrenaline starts flowing? Our ego and sense of self-worth are closely tied to the amount of money we earn, and we are quick to make a (sometimes subconscious) evaluation of whether we are earning more or less than the next person. Some of your customers, therefore, may react emotionally when the subject of rates is brought up. Beyond the issue of self-worth, the project on which you have been asked to consult may be experiencing severe technical or financial difficulties. The client hopes that you will pull him "out of the fire" in a reasonably expedient manner. Your client may feel as vulnerable as I do when I walk into an unfamiliar dentist's office—not sure whether my teeth or my wallet will suffer the most!

IS YOUR MONEY ATTITUDE REALISTIC?

Realistic attitudes about billing rates and the value of your time can save a lot of heartache and let you get on with your business. Do not be concerned with charging top dollar when you start out. Your intent is to establish yourself as a vendor of good service, not as a pirate on the high seas! If your clients are mature, they will understand that you also have expenses and overhead, and they won't accuse you of taking advantage of them.

Many consultants revise their rates every January and adhere to a single rate valid for the entire calendar year.[1] It is good business policy to charge *all* customers the same rate and not to haggle with them on this matter. Why? People in the industry talk to one another. If you are doing a job for one company at $60 per hour and are quoting work at another company in the area at

---

[1] Rates will vary from year to year, reflecting current market conditions and your growth on the "expertise curve."

$100 per hour, mutual acquaintances at the two companies will eventually find out! By lowering your hourly rate for one client, you are setting a precedent for the others to ask for similar treatment.

As with all policies, I find that implementation is more difficult than statement. That is, I occasionally find myself charging certain customers slightly different rates at any one time. Simply setting a rate and sticking with it for a calendar year doesn't always correspond to contractual practicality or business opportunity. For example, if a contract starts on September 1 and carries through to January 31, I would probably quote the whole job at a single rate. Then, in the month of January, I would be working for this client at the previous year's rate, but other contracts signed and performed in January would be billed at the new year's rate.

There are other situations in which I sometimes adjust my rates:

1. When a client buys a significant block of my time in advance, I propose my standard rate. If the customer accepts, everything is fine. If they feel that they should be getting a better price to compensate for concessions made on their part, I drop my standard rate by 20 percent and label it a "long-term" rate.

2. If I must appear in court as an *expert witness,* I charge a significantly higher rate, typically double my usual hourly rate.

3. If I have been hired to do a four-hour training session, basing the fee on the four hours does not reflect the time required to prepare. (As a rule of thumb, each hour of classroom time takes six hours of preparation.) Most consultants therefore quote training sessions on a per-job basis.

4. When a client has an "emergency" that requires me to work weekends or nights, cancel personal plans, or otherwise inconvenience me, I charge 50 percent more than my standard rate.

YOU ARE NOT SELLING ORANGES

When companies need a tangible product such as a pump or a valve, they can easily shop for it on the basis of price. By looking in catalogs and stores, the buyer can get a good idea of the going rate. When it comes to services, the situation is different. It's harder to put a price tag on a service because *what you get* depends greatly on who is performing the service. Many people pay 50 percent more to visit a hairstylist instead of a barber. They gladly pay the extra amount, for the haircut they get from a barber is not the same haircut they will get from their stylist. Therefore, in your sales approach, emphasize quality and value to the customer.

Never advertise on the basis of "low rates" or "reasonable rates." You are not selling oranges at a fruit stand! Your commodity is a much scarcer item, and you want to project the idea that what you have to offer is, in fact, *valuable.* Therefore, do not counter this thought with the idea that you are giving out bargain prices or meeting the competition's rates.

> This is my rate and I'm worth every penny!

I cannot recall a single case in my own consulting experience where the mention of low hourly rate contributed to winning the job. The overwhelming reason for my winning contracts was my clients' expectation that I could *solve their problem.* All other issues pale in comparison.

If you desperately need additional work, it is better to lowball (i.e., bid low on) a fixed-price job than take time-and-materials (T & M) basis work at a significantly lower hourly rate. You will be working for a low effective hourly wage, but you will not be setting a *precedent* that makes that low rate explicit to your customers! (Keep in mind, though, that many small companies have gone out of business by being legally constrained to perform a contract that was lowballed.)

ON NEGOTIATING BILLING RATES

You are in a sales meeting with a potential client. After the exchange of pleasantries, she asks: "What is your rate?"

How do you respond? You haven't had a chance to find out what her problem is, and she hasn't seen your technical presentation. This situation occurs commonly. When it happens to me, I just tell the person my rate. If she goes on to other matters, fine. If she responds, "Well, that seems quite high; we

---

 Tip

In sales meetings, always postpone discussion of your charge rate until you have uncovered the client's need and shown how you can satisfy that need (i.e., solve the problem). Clients look at money differently when they perceive they are buying *solutions to their problems,* not *x* hours of a consultant's time.

have two fellows who help us out occasionally at half that rate," you need to gather more information.

You could respond with "I think I can show you how it would be cost-effective to use me in this situation, even though my hourly rate is higher." If she is still not interested in getting on to the technical portion of your meeting and indicates that the only way she can do business with you is for you to slash your rate, I recommend you leave quickly.

Assuming that your rate is in line with industry levels, you do not need to sell your expertise at "distress" prices. If a prospective client has a distorted view of consulting rates, that is *her* problem. The great majority of businesses that use consultants do not bargain in ways that deprive you of your basic dignity. If you come across an exception, it is better not to do business with that client in the first place.

I do not wish to give the impression that you should be totally rigid in negotiating your hourly rates. For example, suppose you have quoted a rate of $105 per hour for a job. The client responds, "We have a policy that consultants with rates over $100 per hour must be approved by the president, who is away for a month. Would you take $95 per hour to get going on this urgent problem?" I would. There are many other mechanisms that have the same effect. It is reasonable to expect small variations in your billing rate, but don't allow *large* variations due to price slashing, "favored customer" pricing, or loss leaders.

A loss leader situation is one in which the client says, "You're not a proven quantity to us. I'll give you the contract, but charge us only half of your regular hourly rate for the first six months. After that, we'll pay your regular rate." Fellow consultants tell me that the dangled carrot rarely materializes. In practice, the contract is never renewed at the higher rate when the six months are up. At that point, you can either leave or continue to work at the loss-leader rate indefinitely. It's similar to the situation where a boss says to a direct employee who has performed exceptionally well in the past year, "If you continue to do a good job, I *might* promote you next year." Recognize loss leaders for what they are: manipulations to get you to accept less than you are worth.

I have occasionally tried to turn an offered loss-leader deal around so that it would be advantageous to me. I would explain that I was *not* an unproven quantity, that my credentials and reputation were ample to warrant paying my standard rate. Most of the time, these efforts have failed. In retrospect, the very clients who try this are exploiters who have no intention of paying a reasonable rate to anyone in the first place. They will methodically seek out consultant after consultant until they find one foolish or desperate enough to fall for their line!

Tip

Turning away loss-leader work is easy enough when you are busy, but it feels *terrible* when you don't have other potential customers waiting. Therefore, when you're looking for a contract, always deal with *at least* three active prospects so you don't put yourself in that "either this lousy deal—or nothing" dilemma.

In closing this section, I would like to give a personal reason why I don't haggle over rates. As a professional, I *choose* not to focus on rates. Most people don't argue with a doctor or a lawyer about rates; the issues are those of competence and cost-effectiveness. I feel a lot better setting a reasonable rate and sticking to it than considering it as totally negotiable. I want to have my clients pick me because they feel I can do the job well and efficiently, not because I am the lowest bidder on the block!

HOW MUCH TO CHARGE?

There are two basic methods for arriving at a billing rate for your services:

1. Calculate it based on your determination of salary, overhead, fringe benefits, and administrative costs.
2. Determine what the market will bear for your level of expertise.

Both methods are approximate. The greatest uncertainty in Method 1 is the prediction of how many billable hours you will accumulate in a year (i.e., your gross business level). Method 2 is approximate because the market fluctuates and because you may not correspond exactly to a given level in a table describing degrees of expertise.

METHOD 1: DETAILED CALCULATION OF COSTS

Some branches of the government require breakdowns of the billing rate as part of the proposal package. In such cases, it is best to use Method 2 and justify it with Method 1 calculations adjusted to bear them out.

Table 3 shows a breakdown of hourly billing rate for a typical consultant in individual practice. Before bidding on a proposal requiring this breakdown, you should check to see if there are maximum limits imposed by the

TABLE 3. Sample Billing Rate Breakdown

I.   *Direct labor rate* [DLR] = $35.00 per hour
     (Equivalent yearly salary is $35.00 × 2,080 = $72,800 per year)

II. *Overhead rate* [OR] (based on 1,752 billable hours per year)

   a. *Fringe benefits*   *Yearly cost*
      Medical insurance $3,000
      Retirement fund   10,000
      Vacation 20 × 8 × 35   5,600
      Sick leave 10 × 8 × 35   2,800
      Holidays 11 × 8 × 35   3,080
      Disability/life insurance   600
      Sum of fringe benefits = 25,080/1,752 = $14.32 per hour

   b. *Company operating costs*
      Business insurance   600
      Rent and utilities   6,500
      Stationery and supplies   2,000
      Office equipment   2,500
      Qualified travel   2,500
      Accounting and legal   1,000
      Technical books and journals   1,500
      Marketing   4,000
      Prof. society memberships   500
      Prof. seminar and conferences   1,500
      Sum of operating costs = 22,600/1,752 = $12.90 per hour

   c. *Overhead rate* = a. + b. = $27.22 per hour

III. *General & administrative costs* [G & A] = 15%
        Business management
        Payroll and tax administration
        Contractual administration
        Government security clearance administration

IV. *Profit* [P] = 10%

V.   *Billing rate* = (DLR + OR)(1 + G & A)(1 + P)
                    = (35.00 + 27.22)(1.15)(1.10)
                    = $78.71 per hour

customer on profit and general and administrative (G & A) costs. Studying the figures for typical overhead and G & A categories will convince you that the costs of running a business are very real. Being your own business entity justifies you to allocate some of your billing rate to these important business functions.

TABLE 4. Typical Billing Rates

Level of Expertise	Typical Hourly Billing Rate in 1998 Dollars
Nobel laureate	500
Department head of large university or consultant with international reputation	300
Full professor or nationally known consultant with 20 years' experience and many publications or patents	250
Senior consultant in large consulting firm with 20 years' experience and P.E. license	200
Senior consultant in individual practice with 20 years' experience and P.E. license	150
Junior consultant in large consulting firm with 5 years' experience and M.S. degree	125
Junior consultant in individual practice with 5 years' experience and M.S. degree	75
Senior contractor with 15 years' experience and M.S. degree	60
Junior contractor with 3 years' experience and B.S. degree	40

The billing rate breakdown is a *projection* based on the assumption of full utilization. In Table 3, this is taken as 2,080 hours per year minus the time for vacation, sick leave, and holidays. This works out to 1,752 billable hours per year. (When the actual financial performance is tallied at year's end, revised numbers could be determined for the actual number of billable hours, overhead, etc., thus allowing a retrospective calculation of your equivalent yearly salary. Nevertheless, it is the *projection* for the current year that should be submitted to the customer.)

The billing rate breakdown is illustrated in detail so that you may construct your own table, adding or deleting categories as required. In most proposals requiring breakdowns of the billing rate, it is sufficient to give only the direct labor rate, a lumped overhead rate, G & A, and profit. Please note that the purpose of Table 3 is to explain your rate to your customers. For tax purposes, some of the categories may not be deductible expenses. See your accountant or IRS Publication 334, *Tax Guide for Small Business,* for more details.

The *details* of your overhead rate should not be subject to client approval. If you select a more expensive medical plan, the client has little right to demand that you select a lesser one. Likewise, if you give yourself four weeks' vacation, the client has no right to demand that you charge them for only two weeks because that is what they give *their* people. The point is: It's your company. The only constraint is that you keep your overhead rates in line with industry averages, which are typically 140 percent of the direct labor rate.

In Table 3, the retirement fund figure of $10,000 is based on approximately 15 percent of the equivalent gross salary, which is $72,800 per year in the example. In your own calculations, be sure to use a retirement fund figure that reflects your own equivalent yearly salary.

METHOD 2: WHAT THE MARKET WILL BEAR

Table 4 is a guide to Method 2 in 1998 dollars. In this table, distinction is made at the senior and junior levels between rates charged by consultants practicing individually and those practicing within a large firm. The substantial difference is due to the much-increased overhead and G & A for the large firm.

The last two categories, senior and junior contractor, are intended for comparison of the billing rate only. In this case, the employee of the contract house or jobshop receives about two-thirds of the indicated rate. The business arrangement for contract engineers is different from that of consultants: the contract engineer (or jobshopper) is usually an employee of the contract house, which is responsible for marketing, invoicing, and administering payroll and taxes. Most contract houses offer their contract employees a limited set of fringe benefits after a trial period.

The contractor categories are included here because contracting is becoming an increasingly popular means by which companies hire engineers, technical writers, computer specialists, mathematicians, and quality control personnel on a temporary basis. Although many contractors are every bit as skilled as consultants (indeed, some will eventually become successful consultants), they usually lack the credentials and marketing skills required to make it on their own.

If you are still unsure about your charge rate after all this discussion, here's a nifty little method to help you.

1. Take your best estimate as to what equivalent direct employees at your skill and responsibility level are earning. For example, say this is 80K per year.

2. For consultants in individual practice, multiply this number by 1.5 to get your ballpark hourly charge rate. For example, $80(1.5) = \$120/hr$.

3. Actually, there is a Gaussian distribution of what different customers will be willing to pay you. Your ballpark number is a statistically central number with which you can live. Now, take your ballpark hourly rate and calculate ±50 percent of it. To continue the above example, these rates are $60/hr. and $180/hr. These two numbers are important guidelines marking the boundaries of *grandiosity* and *self-denial*. (See Figure 5, where the abscissa has been normalized to 100%.) Unless there are mitigating circumstances, when your rate goes below the self-denial boundary, you're not charging enough. And when you exceed the grandiosity boundary, you're charging too much. The idea is to stay in the range of normality.

HOW TO INVOICE YOUR CUSTOMERS

After you have been in business for a while, you will find that printed invoice forms are a great time-saver. You can generate these forms two ways. With the old method, you design an invoice form and have a printer make them for you. Since many clients require invoices in triplicate, you can order the printing on four-part color forms using NCR (No Carbon Required) paper, which

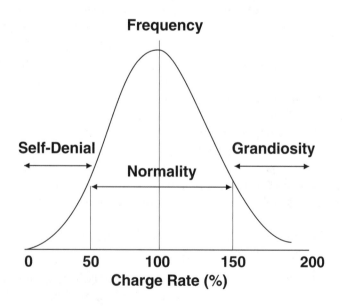

FIGURE 5. Distribution of consulting charge rates.

MEET THE PROS UP CLOSE AND PERSONAL

Alison Balter is the president of Marina Consulting Group in Westlake Village, California. Her company offers training and consulting in database development using Microsoft products such as Access and Visual Basic. Alison is the author of *Mastering Access 97 Development,* pub-  lished by SAMS, and has made over 100 computer training videos for Keystone Learning Systems Corporation that are internationally distributed. She has fourteen years' experience in the computer industry and is a Microsoft Certified Professional.

Alison started her consulting business in 1990. In 1993 her husband, Dan, joined the company. By 1994 her business was booming. Customers were eager for her help, and Alison took on every contract that came her way. She charged $50 per hour. By 1996, thanks to her books and seminars, she had achieved greater status as an expert and was able to charge $100 per hour and $800 per one-day seminar. Her schedule was full for three months in advance.

By 1997 Alison was still booked solid three months in advance. Customers were lining up outside the door and she was working more hours than she wanted to. One day Alison was casually discussing her dilemma with a client who was an experienced lawyer. He told her, "You're booked solid for the next three months, right? Well, that means you're not charging enough. Raise your rates until you start to get a few open slots in your calendar."

Alison wondered, though, "Will my customers walk away if I raise my rates even higher? Will they think I've become too cocky?" She reluctantly raised her rates to $150 per hour and $1,500 per one-day seminar. To her amazement, her clients didn't even question the raise. These large companies could afford her rate. They wanted her, and no one else would do.

With the higher billing rate, Alison can afford to turn away business that does not align with her goals and works only fifty hours per week. She is now able to spend more time with her family and enjoy the fruits of her efforts.

Alison is one of the most successful and dynamic consultants you will ever meet. She says, "The best thing about consulting is the relationships I have developed. I'm always meeting new people, making new friends. It's ironic — I'm supposed to be the teacher, but I feel like I'm the one who is the learner. It's exciting to be paid to learn about a client's business and the stories they share."

eliminates the need for duplicating. You fill in the forms using a typewriter or a dot matrix printer that has multiform capability.

The new method simply keeps the form inside your computer. You design the form once. Then, as required, you fill in the form on your computer screen. Many business accounting programs have invoice modules that allow you to generate as many copies of an invoice as you need. When laser printed, these invoices look very professional.

Example 15 shows my own design. It includes small boxes for the following: invoice number; date; terms; client's purchase order number; time period of work; client name and address; hourly charge rate; service, material and expenses; and total amount of invoice. The main body of the 8½-by-11-inch form has space for tasks performed, hours worked, and price subtotal. There is also a space for my signature at the bottom, as many companies require this for their records. At the top of the form, the word INVOICE is written in prominent letters. My company name, address, and telephone number are given immediately below.

The layout of your invoice may be somewhat different, but allow enough flexibility to adapt to future changes in your business. For example, if there is any possibility that you will be billing your clients for the services of an assistant or an associate in the future, do not print "for the services of John Smith" on your invoice form. Instead, print "for the services of _____."

Always keep a copy of each invoice you send out, preferably the same color if you are using four-part color forms.

GETTING PAID

Seventy percent of all bill collection problems can be prevented by taking some precautions before you ever send the bill.

1. Make sure your payment terms are explicitly stated in your proposal. For example, "Terms are net thirty days."
2. Make sure the client's purchase order reflects payment terms that are agreeable to you (in the event that they are not *identical* to the ones you proposed). If not, call the purchase agent *before* signing the acceptance copy of the purchase order.
3. Address your invoice to the proper person and department. If you send it to the wrong person, it may be delayed or lost.
4. Call it an INVOICE. Don't use misleading titles for your bills, such as "Statement" or "Professional Services Rendered." Your billing form should have the word INVOICE written in large letters at the top of the page.

INVOICE

HK   Harvey Kaye & Company
42 Bent Oak Trail
Fairport, NY 14450

716-223-4502   **Consulting in Mechanical Engineering**

To:	Invoice Number
	Date
	Your Order Number
Terms	Time Period of

For services of: @ $ /HR

Task Description	Hours
Total Hours	
Subtotal	
Material Costs:	
Expenses:	
TOTAL AMOUNT THIS INVOICE	
Signed by:	

EXAMPLE 15. Sample invoice form.

5. Don't predate your invoice. If you send it on the thirty-first of the month —with the date on the form filled in as the seventeenth—don't expect to receive payment two weeks early. Most clients' accounts payable departments are careful to date-stamp all incoming invoices. The clock starts ticking when they *receive* the bill, not when you send it.

6. Be sure to put your *invoice number* on the form. This way, the client can tell this month's bill from last month's. Your very first invoice should be number 101. Keep the numbers in sequence based on date of invoicing. (Accounting software packages will do this for you automatically.)

7. Make sure your address is on the invoice form so the client can send you the check. Put your phone number on the form so they can call you about difficulties to be resolved.

8. Make sure your arithmetic is correct!

9. Make sure you reference their purchase order (or consulting agreement) number. Without it, some large companies won't know how to forward your bill to the correct internal group for approval.

10. Keep accurate records. Make sure you have a log that details all expenses and time spent on the client's behalf. You may need this log in the event of disputes.

11. Invoice your clients promptly. Don't put off this essential chore because you are too tired or busy.

12. Briefly describe the work performed during the billing period. Unless your contract specifies otherwise, a phrase or two is usually sufficient.

13. If you are concerned about a client's ability to pay, suggest a shorter billing cycle in your proposal. This will limit your financial risk. Billing cycles of every two weeks (and even weekly) are not uncommon. Another tactic is to ask for advance payments in the proposal. This is especially easy to justify if you must make out-of-pocket expenditures at the beginning of a project.

14. When it comes to small companies, whether you get paid often depends on how well the company is doing that month and how the CEO views your contribution toward the company's welfare. To avoid billing problems with small companies, stay in constant touch. Establish good rapport with the CEO to know whether she is pleased with your work.

15. Do excellent work! One way to ensure that clients pay is to provide services that are competent and professionally rendered. Issue progress reports and interim results to remind the client of your presence and good efforts.

When a client goes bankrupt, it will be very difficult to collect any consulting fees that are owed to you. There is an established order that bankruptcy courts follow in paying off obligations of bankrupt companies: lawyers receive payment first, secured lenders second, taxes are paid third, unsecured lenders receive payment fourth, and general creditors — such as consultants — last. Moreover, if you take the client to court, it might take years and thousands of dollars in lawyer's fees to claim the few pennies left over for you. Your best bet is to simply forget about collecting the debt; write it off as a business loss on your taxes.

Of course, the best cure to this bankruptcy ailment is an ounce of prevention. If your client seems on financially shaky ground — or you don't under-

---

SIGNALS THAT A CLIENT IS ABOUT TO GO BELLY-UP

- They are slow in paying your bills.
- They lose a major customer.
- They are being sued for a major liability case.
- They start laying off or firing staff.
- The principals are always away from the office. (They are arranging a "distress sale" to pump up business volume or desperately trying to sell the business before it goes under.)

stand how they can afford to do business with their current customer base and overhead— chances are they are flirting with insolvency. Your ounce of prevention consists of *insisting* on weekly billing and never letting your unpaid balance reach more than a month's work. Companies in this precarious situation will counter by asking you to bill *less* frequently, claiming that it upsets their accounting system. Resist! Don't be bullied into submitting invoices less frequently than you feel comfortable with.

What happens if a customer does not pay on time? Wait ten business days beyond the due date. If the check has still not appeared, call their accounts payable department and ask about the payment status of your invoice. If it is in the payment queue, ask for the expected date of payment. Write the name of the clerk in your telephone log. If it is not in the payment queue, try to determine if they have received the bill, and whether it is still in the sign-off (approval) stage.

If payment is delayed by more than a month and the accounting clerk gives you excuses every time you call, you have a problem. The first step to resolving it is to call your client's *technical* contact and ask if there is a problem with your bill.

If indeed the client does not want to pay or is having financial difficulties, you are faced with a serious challenge. On the one hand, you can threaten to stop work on that project—or actually stop work—until the client pays up. This is sure to cause friction.

---

[2]Second notices embellish the original invoice with warnings such as "SECOND NOTICE. Your account is now 30 days past due. Please remit the amount due IMMEDIATELY."

Invoice the client again, marking SECOND NOTICE on the form.[2] Make every effort to discover the root of the problem before another two weeks go by. If the client appears happy with your work and seems willing to pay you but the company has problems cutting you a check, ask to meet with the manager who approved your contract in the first place. In the meeting, emphasize that if you are not paid, you cannot meet your financial obligations, which would seriously impact your ability to deliver services to them. Then wait a week. During this time, quietly gather up any critical work you have done (and which they don't "own," since they haven't paid you). Yes, prepare to run off with the goods. At the end of the week, if no payment is received, find any good excuse to stop work temporarily, even if the client is in the middle of a do-or-die panic. Take as many of *your* goods with you as you can. These are your "blackmail chips."

You have every legal right to withhold data and information that you have developed in the course of your consulting until the client pays for it. The client will not like this and may threaten you with legal action. But the time for good manners has passed.

If you sense a lack of good faith on the client's part or total dissatisfaction with your work, terminate the relationship. If you have a good case, you can take the client to court. Be aware, though, that in most situations, lawyer's fees and lost time will eat up the proceeds. (Collection agencies charge a smaller fee but are usually ineffective in collecting consultant's fees.)

Although other consultants have told me horror stories about bill collection, I have rarely encountered problems with it myself. Most of my clients try very hard to pay on time. After all, *their* reputation is at stake as well. In twenty-five years, I have sent second notices on only five occasions. Only once did I go beyond this: Twenty years ago, a client owed me a substantial amount, and I was running into a cash flow problem. There was no response to my second notice. Instead of issuing threats, I made a personal visit to the accounts payable clerk. After we discussed the problem, I showed the clerk a photograph of my young son and mumbled something like "Baby needs new shoes." The next day, she hand delivered my check!

CHAPTER 10

Projecting a Professional Image

CLIENT RELATIONS WILL MAKE YOU OR BREAK YOU

The consultant's goal in client relations is to behave *professionally* and *confidently*. Medical doctors take formal training to develop their bedside manner and the ability to handle troublesome patients, cranks, and emergency situations. They know the trade-offs involved in various doctor-patient situations *before* they occur. Most engineering and technical consultants are not prepared in these techniques, and they are often unaware of their attitudes and mannerisms in dealing with clients.

PUTTING THE CLIENT AT EASE

A large part of client relations centers around *putting the client at ease.* When the client is at ease with you, the groundwork for a strong and trusting relationship is established. This positive relationship will keep your clients coming back again and again. Presenting an acceptable visual image, being easy to work with, doing competent work, and being considerate of the client's interests all contribute to the client's good feelings about your presence. These aspects will be discussed shortly.

PUTTING YOURSELF AT EASE

The other, less obvious part of client relations is *putting yourself at ease.* This is accomplished by feeling worthy and having the confidence to deal with unrealistic client expectations. Feelings of self-worth are difficult to nourish in the consultant's environment, for many of the clients' personnel are overcome by the negative thinking described in Chapter 1. These negative thinkers feel powerless to change their situation and have low levels of self-esteem. They may be unable to offer the consultant recognition, "good vibes," or even courtesy. Further, the consultant's presence may be a reminder to the client's personnel that the client's organization is poorly managed, understaffed, or incompetent.

To avoid stumbling over these obstacles, look to yourself for the motivation to do competent work in a professional manner. A positive attitude as described by Denis Waitley in *The Psychology of Winning* and by Anthony Robbins in *Awaken the Giant Within* (see the Suggested Reading List) is an integral part of this motivation. Positive thinking allows you to focus your creative energies on achieving your goals rather than dissipating these energies in brooding over countless negatives.

In proposing work, you should not feel inadequate when a prospective client decides that you are unsuitable, overpriced, or unfamiliar with the particular application. Clients have their own criteria and preferences for selecting consulting help, and being turned down is not a reflection on your personal worth.

HOW TO DRESS: MY "ONE-OVER" METHOD

One of the most important aspects of putting the client at ease is presenting an acceptable visual image. Dressing well is not solely for the client's benefit. You will find that dressing well gives you a noticeable boost in confidence and self-respect.

I have a theory about the way a consultant should dress when visiting a client. It's called the "one-over" method: *Try to dress one level over the client's personnel, but no more than one level over.*

Most engineering companies are at level 2. Therefore, on the average, a two-piece suit would be overdressing. However, if a particular client is at

LEVELS OF DRESS OF CLIENTS' PERSONNEL	
Level	**Description**
0	Overalls (plant engineer)
1	No jacket or tie (shirtsleeve engineer)
2	"Casual" jacket (a category specific to engineers and programmers)
3	"Neat" jacket and tie
4	Two-piece suit
5	Three-piece suit

level 3, I recommend wearing a two-piece suit. When you are asked to visit the client's new process plant where you will be talking with the plant engineers, for maximum credibility you should dress no higher than a level 2.

If in doubt on the first visit to a new client, wear a two-piece suit. This is not the end of the matter, though! Read a good book on executive wardrobe such as *John T. Molloy's New Dress for Success,* cited in the Suggested Reading List. This book will help you focus on wardrobe essentials such as fit, color, pattern coordination, and materials. Cost-saving measures discussed in this book include the advantages of planning and how to avoid poorly made, little-used, or frivolous items.

ON THE IMAGE YOU PROJECT

There are many situations where diametrically opposite advice is given, depending on whom you ask:

He who hesitates is lost.	*Look before you leap.*
Opposites attract.	*Birds of a feather flock together.*

When it comes to consulting relationships, I have found that opposites do *not* attract. That is, it's much easier to build trust and rapport when the consultant and client look similar and act alike, when they share similar values. One corollary of interest here is that *it's difficult to build a relationship with a client if you differ in significant ways:*

1. If your image clashes with the client's expectations of how a consultant *should* appear (unusual haircuts, dangling medallions that advertise nonprofessional affiliations, wild clothes, foul body odor, or strong perfume).

2. If your value systems are very different. For example, you spend your spare time fencing stolen merchandise, playing a tough at the local pool hall, or mutilating your body in an ascetic monastery.

3. If your mannerisms are different. For example, you are a consulting acoustic engineer for a Hollywood studio where "showbiz" language and behavior are acceptable. When you try your Henny Youngman impressions on your non-Hollywood clients, they look at you in disbelief.

4. If your age or gender is very different. For example, you are a sixty-two-year-old software consultant working with a client company where everyone is under the age of thirty. They refer to you as "Grandpa" and automatically discount everything you say as "old-fashioned."

5. If your native language or race is very different. For example, you are a white American consulting for a client company in Beijing, China. On your last visit, you asked a pointed question that embarrassed the client manager. You did not appreciate that in China, asking the boss such questions is tantamount to a slap in the face. Now the manager wants you off the project.

The point is that to facilitate building a professional relationship with the client, there must be a certain amount of common ground. The client must be able to accept who you are in order for rapport to be built. This does not mean that you must imitate the client to get her business. But if your image *clashes* with that of the client, you must find a way to accommodate this fact and overcome the potential disadvantage. Alternately, you can make an effort to find those clients with whom an image clash is less likely.

It is very hard to fool people into thinking you are something you are not. Clients and friends will be quick to sense whether you are in consulting for your own professional satisfaction, the money, or any other reason. There is no substitute in the consulting business for truly being interested in the activity itself.

Conspicuous consumption is not the way to project a good image or feelings of self-worth. Your clients will feel belittled when you open your six-hundred-dollar leather attaché case to show photographs of your fifty-foot yacht. Riding around in a Rolls Royce will only elicit doubt and envy in your clients. (So leave the Rolls in the garage and use the Olds to visit your customers!)

You can *tell* your clients that you are motivated by professional satisfaction, but if your nonverbal messages do not corroborate your words, most clients will not believe you. Unknowingly, you send out many different nonverbal cues that your clients can pick up. These include eye motions, facial expressions, posture, hand gestures, the tone of your voice, even the way you breathe! Genie Laborde's *Influencing with Integrity* (see the Suggested Reading List) describes how this nonverbal behavior is part of the overall communication process. In her book, she shows how the nonverbal parts of the communication process can be used to achieve the outcomes you desire.

The materials in your briefcase form another kind of nonverbal communication. If issues of *Popular Boating* and *Playboy* are visible, the client who sees them may also be interested in the subjects covered by those magazines. This could lead into a discussion that results in greater rapport with that person. But it is also possible that the client may be turned off by those subjects. I play the percentages and keep only professional materials in my briefcase.

Tip

It is possible to take this process one step further. This same nonverbal form of communication can be used to *advertise* your professional interests and capabilities. I often leave my briefcase open, with the latest technical journal or specialty book in my field visible. You don't have to say a word about it. Your clients will notice and it will register in the right places.

MY LIST OF THINGS *NOT* TO DO IF YOU WANT TO MAINTAIN A PROFESSIONAL IMAGE

Learning to be considerate of your clients' interests is an important part of a professional manner. Toward this end, there are many positive things that can be done. But I am presenting a list of NEVERs because one really bad goof can label you persona non grata. I have seen consultants commit some of these NEVERs and have been privy to comments about their ethics afterward! Your reputation will follow you around. I *guarantee* it.

- NEVER discuss your rate with the client's "troops." Most of them are dissatisfied with their salary and can only look antagonistically at your success.
- NEVER complain. It does no good anyway.
- NEVER act boisterously.
- NEVER knock your client's management. Keep your thoughts to yourself.
- NEVER say anything about yourself that would make your client feel inadequate by obvious comparison.
- NEVER bluff on a technical point.
- NEVER knock your competition.
- NEVER foster romances with your clients' personnel.
- NEVER abuse your clients' proprietary interests. Do not mention details about one client's business to another client. If asked to discuss other clients' work, be vague or evasive as the case may require.
- NEVER goof off on your client's time. Read your newspaper and clip your nails on your own time.

- NEVER take pencils, paper supplies, computer disks, etc., from your client's stockroom for your own personal use. People notice.
- NEVER make important personal[1] phone calls from a client's office. It is not private, and it is improper to use a client's time and money for your own purposes, even if your client "knows" you and says it is all right. It isn't!

DON'T "CHARM" YOUR CLIENTS!

I believe that "personality school" training methods are not particularly helpful to consultants. You all know these courses; they promise popularity and business success. Their basic premise is "Be nice to others and others will be nice to you." These courses are potentially debilitating to the consultant in that they want to *paste on* rather than *build in* genuine appreciation of others. They emphasize the manipulative aspects of relationships: I'm going to be so nice to you that you're going to feel terrible if you're not nice in return.

I don't like the attitude behind these manipulations. Further, I have found that you don't have to toady to a client to win her business or her respect! Give yourself and your customers credit for having a *legitimate* business relationship. Neither one of you needs to coddle or manipulate the other!

Much of the way you conduct yourself will depend on the (partially subconscious) image you have of the consultant-client relationship. My image of the relationship is that both client and consultant are part of a valid and justifiable business relationship. You are not doing the client a favor, and she isn't doing you a favor. Some consultants have the distorted view that they are doing the client a favor. And some clients have equally arrogant postures. You can drop these postures when you realize that there is a more professional way of looking at the interaction.

I again recommend Laborde's *Influencing with Integrity* as a guide to dealing with your clients in an open, nonmanipulative manner.

DOES A SENSE OF HUMOR HELP?

Is a sense of humor essential to an engineering consultant? Many successful consultants are totally serious. However, they develop ulcers and have heart attacks at an early age. A sense of humor may not have a monetary payoff,

---

[1]By "personal," I mean phone calls to other clients or for purposes unrelated to your client's project. Of course, it's OK to make brief calls to family and friends to confirm plans.

<u>Warning</u>

Never tell jokes that reflect poorly on your client's management or mock the client's project. Don't even hint at it! I once consulted on a project dealing with the AAMRAM missile (<u>A</u>dvanced <u>A</u>ir-to-Air <u>M</u>edium <u>R</u>ange <u>M</u>issile). There was a creative young man in the drafting department who made a cartoon of an F-14 fighter jet holding little piglets in the missile bays instead of AAMRAMS. He called them SPAMRAMS (a takeoff on SPAM, the pork-based food product). Most co-workers thought the cartoon was funny, but when the company brass saw it, they were outraged. The young draftsman lost his job on the spot. There is an upside to this story, though. A few months later, the budding artist landed a job as a cartoonist for a major magazine and lived happily ever after.

but it sure helps you face life's tougher moments. Be able to laugh at yourself. Not taking things too seriously is a definite personality asset.

If it agrees with your style, learn *how* and *when* to tell jokes to create a light mood. Accumulate good jokes and anecdotes in a file and practice delivering them with style. Avoid shaggy-dog stories and elaborate anecdotes that try the listener's patience. Consider your audience carefully. Don't use purple jokes or stories that poke fun at ethnic groups.

Humor is more than telling jokes and throwing rubber chickens on the conference table, though. True humor has a spontaneous element that cannot be rehearsed. The resulting laughter has an amazing effect: it releases tension. Dealing with clients is stressful enough for both parties. By allowing the element of humor to occasionally shine through, you will feel more relaxed and confident. For more on the place of humor in the workplace, see the delightful book by C. W. Metcalf and Roma Felible in the Suggested Reading List.

THE ART AND SCIENCE OF MEETINGS

Learning how to hold meetings and participate in them is an important challenge for consultants. Successful participation in meetings can bring you more contracts, greater respect, and enhanced credibility.

Part of your meeting and presentation strategy depends on who has called the meeting and who is to run it. If you called it and intend to run it, then you should have a clear idea of the meeting objectives before you set up details such as time and place. Are you calling the meeting to

- inform the client of your progress?
- find ways to resolve a potential obstacle?
- convince the client of a design decision you already made?
- persuade the client to buy certain materials or equipment?
- resolve an inconsistency between two conflicting specifications?
- propose new work as an extension to your current contract?
- ask for technical input from other groups and individuals?

Once you know your purposes in calling the meeting, you can determine your presentation strategy. Does your session look more like a formal lecture, a sales pitch, a negotiating session, or a technical forum? Do you want to merely inform the attendees or have them participate in a group decision? What *actions* do you want to result from the meeting?

The next step is to prepare a one-page meeting agenda to send to participants. State the purpose of the meeting in twenty-five words or less, and include an outline list of the major topics. The list of attendees should be explicitly mentioned, as well as the place, starting time, and duration. Unless there are extenuating circumstances, invite *everyone* directly involved with the subject matter. Indicate people who *must* attend with an asterisk, and cc: people who are invited to attend for background information.

 Tip

As a general rule, the smaller the number of attendees, the more efficient and focused the meeting. If you invite the "big bosses," expect to be sidetracked to a certain degree, for they often regard *all* such meetings as a vehicle for demonstrating their power—even the power to bring up irrelevant and unimportant issues.

You are now ready to assemble your presentation. Regardless of whether you are working from paper or an electronic slide show, determine the topics that must be discussed and the order in which they are best given. The traditional general-purpose presentation format is:

- Introduce the subject.
- Present background information and data.
- Discuss issues and alternatives.
- Lead toward resolutions, recommendations, and action items.

Once you have assembled your presentation and handouts, take the time to rehearse it, even if you must do so on your own time at home. There is nothing worse than delivering a half-baked presentation. You will feel clumsy and your clients will judge you inept. To improve your delivery, I recommend reading *I Can See You Naked* by Ron Hoff. This excellent and entertaining book on presentations is full of useful techniques and advice.

If you're running the meeting, do your best to keep it on track. This means anticipating a variety of problem situations and knowing how to handle them. For example, the "big boss" may have a habit of coming late to meetings and forcing you to start your presentation over from the beginning. (From his viewpoint, the meeting doesn't start until *he* arrives!)

You have some choices: You could make everyone wait until the big boss arrives. If he is more than fifteen minutes late, you could proceed and hope he never shows up for your meeting at all. You could postpone the meeting to a later time. You could design your presentation with "filler" material in the beginning, so little would be missed if he is only slightly late. Finally, if you know this to be a problem in advance, you might prepare a one-page synopsis labeled "Executive Summary for John Q. BigBoss." When he shows up an hour late, hand him the page and casually mention, "I thought that board meeting might tie you up, so I prepared a summary of our progress for you." Unless he is an absolute ogre, he will let you proceed.

Another annoyance that throws meetings off track is private conversations. The way to handle this problem is simply to clear your throat and loudly (but politely) announce that you'd like to get back to the meeting agenda. Quickly segue into your next topic with a question that focuses the group's attention away from the wayward individuals.

Running effective meetings with individuals as "intense" as engineers and programmers is like being a traffic cop. Your job is to keep the traffic flowing as you work your way through the agenda. It is analogous to keeping the reckless drivers inside the double yellow lines; you must prevent the more aggressive and voluble individuals from monopolizing the meeting or taking over with their own private agendas. As a traffic cop coaxes reluctant drivers to move faster, you must encourage the quiet or shy attendees to volunteer information and participate.

Here are some guidelines that may help you with your presentations:

MEETING DOS AND DON'TS

1. <u>Do</u> work from an agenda. An agenda gives the meeting structure and prevents it from degenerating into a bull session or argument. It makes you look more organized and on top of things.

2. <u>Don't</u> embarrass anyone in the meeting or show anyone to be lacking. You'll regret it dearly if you pull this stunt.

3. <u>Do</u> avoid distractions by design. If you know in advance that a certain time, place, or condition will lead to distractions, then set up the meeting for a different time, place, or condition.

4. <u>Don't</u> take anyone unawares! Meetings are not the place to surprise your client with new discoveries, especially negative ones. If you are the one initiating the meeting, discuss your meeting agenda with the appropriate client personnel and get their input *before* you send the meeting notice out.

5. <u>Do</u> have other people contribute parts of your meeting. This helps them "buy into" your meeting and makes you the de facto "organizer," which gives you more credibility in the client's eyes.

6. <u>Don't</u> let the meeting drag on. Define its length in advance and stick to it. If there is too much material to handle in a single meeting, break it up into a number of separate meetings.

GOOD REPORTS BECOME YOUR CALLING CARDS

Reports are an excellent way to *manage your image* at the client's company. By submitting reports that are well written, authoritative, and visually attractive, you can enhance your credibility. You can also use interim reports to increase your visibility at the client's company and to highlight potential problems. I recall how one report I wrote for a client circulated from engineer to engineer in the company. Because I had placed my company name and logo at the bottom of each page, it was free advertising. As a result, an engineer in the company whom I had never met called and asked me to propose on a project he was coordinating.

In final reports, you can sing your own praises and emphasize how you delivered the benefits that you promised in the proposal. You can also use a "Recommendations for Further Work" section as a marketing tool to obtain follow-up contracts. Treat this part of the report as a post hoc proposal. Offer persuasive arguments why the client should implement your recommendations and have *you* perform the work. This is a low-cost way to get follow-up work, because you have already done the "research" to determine the proposed follow-up *at the client's expense.* Moreover, you will have a large competitive advantage because other consultants won't have your degree of insight into the problem by virtue of your previous efforts.

If the thought of writing reports and proposals seems intimidating, it's time to brush up on the basics of technical writing. The books by John M. Lannon and Theodore A. Rees Cheney in the Suggested Reading List provide an excellent starting point for self-study.

Practice is half the trick to developing writing speed and ability. The other half is getting feedback from people who are knowledgeable enough to critique your work. Sometimes your customers provide this feedback ("Fix this report! We can't understand what you're trying to say!"). At other times, family members and fellow consultants may be able to help.

To avoid writer's block, don't write a report in one sitting. Break it up into smaller tasks. First gather your data and information. Second, make an outline of everything that must be discussed. Third, flesh out the outline as fast as you can. Just blurt it out. Don't worry about grammar and spelling. Nail the ideas down. Then put your report aside for a while. When you come back to it, check the sequence of your ideas. Is it logical? Are your conclusions supported by the evidence you have offered? Have you clarified subtle points about which the client may be confused? Does the report flow? Does it take the client from point A to point B to point C and then to your conclusions and recommendations?

Tip

Save good examples of the kind of reports you need to write as you come across them in your reading. Use them as models for your own efforts. I keep a file of "Good Writing Samples" for exactly this purpose.

The final step is to check your grammar and spelling. Computer word processors make this easy, but you should still proofread the final report before submitting it. Spell checkers won't catch typographic errors where a word is spelled correctly but is not the word you intended. (For example, both *their* and *there* are spelled correctly, but it is easy to mistake one for the other when you are writing fast.) Here are some additional guidelines to improve your writing style:

- Consider your audience.
- Don't knock or judge the client.
- Cleanse your language of racial, gender, and religious slurs.
- Be brief. Distill the essentials from the superfluous details. Place the latter in an appendix.
- The first page of your report should be a one-page "Executive Summary."
- Avoid long and complex sentences.
- Vary your word choice and sentence structure to avoid boring your readers. Most word processors have thesaurus functions that make this easy.
- Emphasize your major points with bullets or larger type.
- Use graphs, pictures, and tables to organize and present your data.
- Avoid overusing the passive voice ("The report was written"). Use the active voice instead ("John wrote the report").

CHAPTER 11

Dealing with the Client

BUILDING THE CLIENT RELATIONSHIP

Many companies treat their direct employees like family. This metaphor likens the boss to a beneficent parent and the employee to the dutiful and aspiring child. This relationship is not one of equals, but one in which the employee is undeniably subordinate. The boss tells you what to do and when to do it. The company "feeds" you every week with a paycheck. Once a year, at the company picnic, the CEO stands up and insists that the company cares. On your annual review (birthday), the company gives you presents—if you perform an outstanding job. And, finally, they shield you from things they feel you don't need to know (like how much the CEO earns and how much the company spent to recruit that new VP of engineering).

Psychologically, the relationship between consultant and client doesn't fit the family model very well. When you are a consultant, if you think of your clients as parents, you're going to end up confused and disappointed! In fact, allow me to make a crude analogy: The relationship is much more like two people dating. The longevity of the relationship is much less assured. And, exactly like dating, although both parties are "equal," they come to the interaction with different needs and expectations. Consulting relationships are often more fragile than direct employment, but they also offer delicious possibilities that will never occur if you sit in your parents' living room watching TV!

Like any relationship, the consultant-client relationship is enhanced when both parties possess relationship skills. Clients will come back again and again when they *feel* that you consider and respect their best interests. They will show appreciation when you are able to offer good value *from their point of view*. Similarly, *you* will look forward to working with clients when they treat you with dignity and respect. When clients are open and fair with you, you will feel better about giving them that extra bit of effort. The basic components of solid relationships are:

- *Respect:* Concern for the well-being and integrity of the other party. Not merely speaking with deference, but *acting* with the client's best interests in mind.

- *Openness:* The open communication must proceed in two directions: being able to express yourself *and* being able to listen to the other party. When a client says something the consultant doesn't want to hear, many consultants "stonewall" the client, making him angry and resentful. In contrast, with open communication, consultants find it much easier to build consensus with the client on project goals and decisions.

- *Mutual benefits:* The benefits are rarely equal, so in practice this means assuring that each side gets enough of what it wants to feel satisfied with the deal. From the consultant's point of view, this means: *Give them more than they asked for.* Customers like the baker's dozen, the bonus thrown in for good measure. Always provide a small "extra" to make your client's purchase that much more attractive.

- *Ability to resolve differences:* When you and the client agree on issues, life is easy. Your ultimate success as a consultant, however, is determined by knowing how to resolve conflicts. To reach effective compromises, work on your diplomacy and communication skills. In handling disputes, go the extra mile and give the benefit of the doubt (as long as it doesn't mean giving the shop away).

- *Emotional competence:* When both sides act with emotional maturity, they refrain from unfair fighting, cheap insults, hostile humor, and manipulative tricks. Instead, they use patience, compassion, and honesty to build trust into the relationship.

Of course, you cannot control how your clients act, but just controlling your own actions and attitudes in these regards will give you a great advantage over many other consultants who do not take the time to learn or develop relationship skills.

In any situation where a skilled professional helps others, certain relationship issues must be dealt with before useful work can commence. The world's first therapist — Sigmund Freud — was quick to notice that problems usually arose if three specific issues were *not* addressed. The first issue was that of *resistance,* where clients seemed more interested in hindering the progress of their own therapy than in getting their problem solved. The second issue was that of *dependence,* where clients would try to get Sigmund to make their decisions *for* them. And the third issue was that of *control,* where clients would insist that paying the fee meant they were the "boss."

Although these issues affect the consultant-client relationship in different ways, the similarities are very strong: both therapists and consultants must

MEET THE PROS UP CLOSE AND PERSONAL

Bob Kirchhoff is a professor of mechanical engineering at the University of Massachusetts who specializes in fluid mechanics. He has over thirty years' experience in the field, has written over fifty journal articles, and authored a monograph, "Potential Flows: Computer Graphics." He is also the creator of internationally acclaimed video instruction courses on advanced topics in fluid mechanics.

Bob started consulting for a large Fortune 500 company back in 1983 and has been their fluid mechanics consultant ever since. When I asked him what he did to keep them coming back again and again, Bob replied, "You know, Harvey, I was listening to a National Public Radio broadcast where a lady banker was explaining the '4 Fs' of business success. I tried them for myself and they really work. It's simple. The 4 Fs are: Fast, Friendly, Focused, and Flexible.

"First, in dealing with this client, I am always fast in responding. Not just with an answer, but with a complete written report they can use for reference. They feel they're getting more for their money. Second, I try to be friendly by never putting anyone on the spot. I don't find blame or pass judgment, but always strive to acknowledge the positive contributions of the individual contributors. I am open to their ideas, and listen attentively and sincerely. I strive for open communications and include everyone who should be involved. Further, I maintain financial friendliness by keeping my charge rate modest on purpose. I already have a steady income at the university, so I don't need to charge the client top dollar.

"Third, I focus on the client's problem and am not distracted by my own agendas or pet theories. I dig carefully to determine the client's problem objectives and ask enough questions to make sure I'm solving the right problem. Also, I have learned not to be distracted by the client's occasional inability to focus. Especially in working with the client's technicians, I must be very firm at times when they do not appreciate the need for modifying their customary measurement procedures. In a friendly way, I explain why we must do things differently in certain instances.

"Fourth, and most difficult for me, is flexibility. When working with an industrial client, changes can come really fast. The technical parameters governing a problem can be completely redefined overnight.

(continued)

> I have learned to change horses in midstream, to flow along with the changes rather than resist them."
>
> Bob has kept this client through four waves of department management in fourteen years. "Remember the 4 Fs," he says, "and you'll keep 'em coming back just like me."

tread a fine line to provide useful service to the client on the one hand, and protect their own rights on the other. In technical consulting, these issues are usually less pronounced, but they are there nonetheless. Practicing psychologists take years of training to understand the convoluted mechanics of client relations so as to be able to help clients without hurting themselves! The next three sections contain an overview of the same concerns—resistance, dependence, and control—for technical consultants.

RESISTANCE

One major aspect of client relations is developing the ability to sense when the client is *resisting*. Unless directly countered, resistance tends to grow. Nip resistance in the bud, before it swells to unmanageable proportions. Sometimes the resistance clients offer is well articulated and unambiguous. At other times the messages are subtle and may not even be verbalized. Resistance can take many forms. The client may

- reject whatever you say;
- attack your credentials;
- be silent and uncommunicative (stonewall you);
- refuse to respond to your calls and messages;
- confuse the issues;
- string you along about when the project will really start and how much is in it for you;
- allow numerous interruptions to ruin your meetings;
- press you for on-the-spot answers;
- structure questions at technical meetings so that he appears to be the expert and you appear to be the student;
- attempt to "legislate" the solution;
- want too many details or explanations;
- give you too many details (swamp you with information);
- withhold significant data;

- emphasize your shortcomings or criticize your position as a consultant ("Boy, you have it easy. All you have to do is design the thing. Here in the *real world* we have to make the damn thing work.").

The issues underlying resistance are power and vulnerability. The client may view the problem under discussion as a pain, an embarrassment, or a crucial threat to the company's existence. The consultant, on the other hand, looks at it as a way to make some money by helping the client get rid of the problem. The consultant's very presence at the meeting may serve only to remind the client that someone in his organization has goofed or that a competitor has strong-armed him into a do-or-die dilemma.

To deal with resistance, first recognize the form it is taking. Then state to the client—in a nonblaming, neutral, and diplomatic manner—the kind of resistance you are experiencing. Doing this interrupts the client's pattern. Next provide graceful egress for the client by tactfully asking for clarification. Then listen.

For example, suppose a client is asking you many pointed questions about the theoretical aspects of a product's operation. You realize that these questions are irrelevant to the situation at hand. It seems that the client is trying to administer some sort of exam to probe your expertise. You wonder if he is searching for a small hole in your knowledge base to diminish your credibility. You could counter this ploy by saying, in a calm voice, "You are asking a lot of questions about fundamental principles. Do you have any reservations about my ability to handle them? If not, I would like to proceed to the focal issues in this study." Nine times out of ten, the client will stop the interrogation and let you get on with the meeting. Sometimes the only way to handle resistance is to be very assertive.

DEPENDENCE

At the beginning of the project, you want the client to be dependent on you. Near the end of the project, you want them to be more independent so that when you are gone, they will be able to carry on without you.

This is analogous to the therapeutic situation, where the therapist tries to end the therapy on a positive note. For the therapist to merely pronounce, "You're cured" is not enough. The therapist must convince the client that they are indeed finished by virtue of their newfound ability to "do it" on their own. Similarly, the consultant must judge

- what the right time is to terminate the consultant's interaction (as in therapy, the consultant's perception of when the project is over may be vastly different from the client's);
- how to "wean" the client from the consultant;
- how to demonstrate the client's ability to carry on after the consultant is gone;

Part of the determination hinges on your particular situation: Do you *want* to make the client dependent on your services every time the need arises, or do you want to gradually wean them to self-sufficiency? Do you want to do a "short-term therapy" or a long-term interaction? Do you want to hand off the work to a person on the client's staff or to a worker on *your* staff? How do you view your function at the client's place? Are you an adviser, a hired hand, or a trusted team member?

The consultant's general strategy on the dependence issue is always the same: *Strive to achieve and maintain the proper distance.* With the answers to the above questions, you will be able to formulate the details of your strategy. If you maintain the proper distance, your client won't become so clingy that they try to hire you as a subordinate or so detached that they never need your services.

CONTROL

The issue of control usually doesn't get too sticky until you start to perform a substantial amount of work at the client's office. At that point, you may be assigned a temporary office at the client's place. Much of your daily interaction with the client focuses on the one person administering your contract, who, in effect, becomes a pseudo boss. At first, she is very polite and asks for your technical input on important issues. A month later, she doesn't ask; she simply asserts, "I'm in charge here. Let's do it my way." Two months later, she starts to micromanage you; she rearranges your schedule on a daily basis and tells you how every small task must be done. Wow! It's starting to look like direct employment again—only this time, your boss is ten years younger and less experienced. Under such circumstances, your effectiveness is slipping out of your own hands. In short, you are losing *control.*

When you work as an employee for a corporation, you have a certain amount of *position power.* By virtue of your title, you are given authority—control—commensurate with that rank. When you are at the bottom of the corporation, you feel powerless because everyone outranks you. As you advance up the ladder, you feel better, not quite as powerless, because now *you* outrank some underlings.

The situation is fundamentally different for the consultant. When you come into a company, you have no delegated power. You lie completely outside the company hierarchy. Even though you may be a Nobel laureate, as a consultant you have less power than the company janitor. Without power in the client's organization, you will find it difficult to request necessary resources, call meetings, negotiate changes, and even to get past a manager's secretary to discuss important situations. Without power, you may not have open access to significant people and resources. Without power, some individuals may not even recognize your existence. In other words, you don't count!

What employees accomplish through position power the consultant must accomplish through goodwill and political alignment. Yes, by allying yourself with sources of power within the organization, you gain power *by association.* This is why management consultants usually deal only with the top executives in a company. Without top management's strong backing, their efforts will be compromised by infighting at the middle-management level.

The practical solution to the control issue is to make sure you come in through the highest level of management you can arrange. Once you are in, establish a continuing and candid dialogue with them. As a technical consultant, you may not need to deal with top management, since your work is more detailed and less bottom-line oriented than the management consultant's. Nevertheless, exert every effort to meet the highest level of management that is directly involved in making decisions about your funding and work scope.

Please don't get me wrong. I am not a power or control freak. I work well with clients at all levels of the hierarchy and have no problem in dealing with authority. The point is to have enough control to do your job effectively and to maintain your self-esteem. When a client manipulates to deprive you of these, you must countermanipulate to gain them back. Although I strive to be as nonmanipulative as possible in my relationships, sometimes you have no choice but to react to the client's manipulations.

If you cannot find a reasonable balance of control in working with a client; if they insist on treating you poorly and your efforts to regain control are in vain—then terminate the relationship. Remember, only *you* can say how you feel about a relationship. If the client has seized an unfair amount of control and yet insists that they are treating you fairly, look inward. Assuming that you do not suffer from a grandiose self-image, ask yourself how you feel about the situation. Listen to what your gut tells you.

DEVELOP NEGOTIATION SKILLS

Negotiation skills are essential for dealing effectively with clients. You may not need these skills on an everyday basis, but they are a lifesaver when you and the client disagree on contractual matters, interpretation of work scope, intellectual property rights, and the handling of mistakes that arise during the performance of the contract.

The word *negotiation* elicits positive emotional reactions in some people and negative in others. For individuals brought up in authoritarian households, there was no such thing as negotiation—one simply obeyed. The word *negotiate* meant that Mom or Dad was about to impose their will. As adults, these people become fearful that in the negotiation process, they will be powerless. For individuals brought up in liberal environments, the word *negotiate* meant that Mom or Dad was willing to listen and accept reasonable requests, within a wide range of choices. Negotiation to them means the chance to say what you want, and is eagerly sought out.

For consultants, negotiation should have a neutral connotation. In negotiating with a client, you are neither the victim nor the aggressor, neither the powerless serf nor the intimidating monarch. You are merely trying to find the best terms that you and the client can *mutually* arrive at.

Roger Fisher and William Ury, in *Getting to Yes: Negotiating Agreement Without Giving In,* categorize this style of negotiation as "soft" in contrast with the "hard" style often portrayed in TV "Day in Court" programs. With hard bargaining, the parties are adversaries who distrust each other. It's war! Battle positions are established, and each party seeks to win at the expense of the other. With soft negotiation, the parties are friends who trust each other and whose mutual goal is agreement.

In my opinion, the *only* type of negotiating for consultants is the soft variety. (Unless, of course, it is clear that the relationship is over and you are dealing with gross injustices committed by one side or the other.) If you want repeat business with your clients, soft negotiation helps build trust and good-

will between you. Power-play negotiation tactics may work once or twice, but they quickly make clients resentful and wary.

For the consultant, negotiation is not a bunch of cheap tricks and unethical tactics, but the creative art of finding compromise where it appears that no middle ground exists. It is the art of "getting your way" and, at the same time, convincing the client that he got his way, too. It is the art of get-

CLIENT NEGOTIATIONS IN A NUTSHELL

- Give and take. Always seek win-win outcomes.
- Don't make demands and ultimatums. They usually backfire and always cause ill will.
- The client's problem is *your* problem, but the reverse is not always true.
- Never present a problem without also giving some good alternatives to solve it.
- Negotiate only with those who are empowered to make decisions.
- Don't take unfair advantage; people remember.
- Some things are negotiable, others aren't. Your integrity is one such nonnegotiable.
- Address the *issues,* not the individuals.
- Getting your point across: State the message differently, but never yell the same message louder.
- Seek allies in high places before trying to scale a stone wall.
- Let the client make the first offer; it may be entirely acceptable without modification — or even exceed your expectations.
- Find the common ground (area of mutual benefit), even if you must *create* it.
- Lose your cool, you're a fool. Never, never lose your temper!

ting two parties unequal in strength, needs, and resources to come to an agreement.

In negotiating with a client, remember that your power and goals are different from the client's. You must always consider your relative positions and strengths in carrying out negotiations. You must understand, at least in a general sense, how much you are willing to give in, how much you want to bluff, and what issues are nonnegotiable (take it or leave it).

Negotiation requires you, the consultant, to put on a different hat. You are not dealing with technical facts and methods, but with people. If you have an overwhelming need to be liked and approved, you will probably be weak and vulnerable as a negotiator. Many negotiations require firmness and sticking to your guns. If not, the other party wins. The client's negotiators are usually quick to sense whether you're a "nice" guy, one who will accede rather than create an outright conflict.

Knowing your boundaries and being firm are not the same as playing hardball. Soft negotiators can still be firm, especially over issues that you consider nonnegotiable. Good negotiation skill comes from being able to say no at times, that is, being able to walk away from the bargaining table. In marketing, if you're saying yes all the time, it's highly likely that you are not focusing your business enough. Similarly, in negotiating, saying yes all the time means that you are consistently placing the client's interests ahead of your own. The price of being so agreeable is accepting certain deals that can hurt *you*.

To better understand negotiation, I recommend Chester L. Karrass's classic book, *Give and Take: The Complete Guide to Negotiating Strategies and Tactics.* Karrass provides a broad picture of hard and soft tactics that will help you develop greater awareness of the other party's negotiating style and motives.

IS SATISFACTION GUARANTEED?

Once the consultant has been awarded a contract, there is no guarantee that the customer will be pleased with the results. For example, a consultant may be asked to do a feasibility study on a new concept promoted by a top manager. If the concept proves unworkable, the consultant (not the subordinates) has the unpleasant duty of telling the boss that his concept is unacceptable.

Trying to please a customer can be downright frustrating at times. A classic illustration is the story of how Michelangelo designed a sculpture for the Vatican.

> The pope summoned Michelangelo and commissioned him to do a marble sculpture for the Church. He was not given specific instructions for the theme or size of the work, so he spent the next three years creating a very large and beautiful masterpiece depicting life-size men and women riding on horses. When it was unveiled, the pope didn't care for it. He wouldn't pay Michelangelo until he came up with something that met *his* approval.
>
> Michelangelo was now three years older and wiser, so he quickly created four new designs in scale model. He brought these to the pope, hoping that at least one of them would meet with approval. And, indeed, one, with minor modifications, was satisfactory to the pope. Michelangelo spent the next three years doing the full-size rendition. Unfortunately, during this period, the pope died.
>
> The new pope looked at the sculpture and told Michelangelo, "I don't like it. I can't give you any money until you come up with something that meets *my* approval!"

The moral of this story: Sometimes you are going to lose, even if you try to anticipate the problems. This did not stop Michelangelo from continuing with a successful career, and it would not stop me either. It *would* encourage me to understand that setbacks *will* occur in the consulting business and to accept them as part of the game.

DEALING WITH UNREALISTIC CUSTOMER EXPECTATIONS

It takes patience and confidence to deal with clients who have unrealistic expectations. For example, a client may expect you to be familiar with a particular device that happens to be the only one of its kind in the country and that has been kept under close wraps. All you can do in this situation is inform the client that it will take time to understand the problem in sufficient depth to produce useful results.

Don't feel bad because you don't know the device (system, program, etc.) as well as the client does. The situation is analogous to that of a patient going to a new doctor. Regardless of how well documented the patient's medical history is, a doctor new to the case will conduct his own interview before recommending a treatment. The extensive questionnaires of previous doctors count for nothing; the new doctor will start from scratch. He needs to understand the problem *to his own satisfaction* from the beginning. Trying to bypass this stage by blindly accepting someone else's evaluations is a compromise of professional integrity.

Occasionally, a client may withhold information that you need to perform the project. This is something with which every consultant learns to live. There are many possible motives for withholding information, but you have only limited visibility about the politics of the situation. The reasons certain people on your client's staff refuse to cooperate may remain a mystery.

If you cannot get the needed information through the back door (third parties and unofficial channels), you might be able to develop your own data. Or, find a way to make the information irrelevant. For example, you could do sensitivity studies based on bounding cases. Perhaps these studies will show that the answer is independent of the withheld data. This strategy effectively eliminates the need for it.

Some clients hire a consultant for a "day" and expect that the "day" will be ten or fourteen hours instead of eight. After all, that is how long *they* work themselves! I deal with this by quoting my rate on an hourly basis. If asked to quote on a daily basis, I oblige, with the additional clarification that a "day" consists of eight working hours.

Other clients try to dictate the price on defined-scope, fixed-price jobs. In this situation, the client indicates that his budget allows only a certain amount for the task. The client may well know that the price is a factor of two or three

MEET THE PROS UP CLOSE AND PERSONAL

Steven Morgenstern is a registered professional mechanical engineer who is a principal consultant at the Kinetic Engineering Group in Poulsbo, Washington. He does consulting projects in industrial HVAC systems.

Steven says that dealing with clients can be downright frustrating at times and offers this "client from hell" story: "My most difficult client problem happened on a commercial renovation project. The owner of the project hired an architectural engineering firm to do the makeover, which, in turn, hired me to design the HVAC and plumbing systems. At the time I accepted the contract, it seemed straightforward enough. However, once the contract was under way, the owner started to place significant design restrictions on the type of equipment that I could use. He claimed that his aesthetic considerations dictated my design parameters.

"The owner soon became involved in my day-to-day activities. At first he was simply excited about the project and enthusiastically wanted to learn as much as possible. Although the owner lacked experience in the building trades, HVAC design, and project management, he kept on asking more questions. As the project progressed, he became more proactive and wanted to understand every detail. I knew I was in trouble when the owner hired a non-engineering consultant to check my design. Moreover, without understanding the design parameters or process, he started gathering conflicting advice from vendors. The meddling finally reached the boiling point: the owner was questioning every single design step and preventing me from making any progress.

"I gained an appreciation for those old auto-repair shop signs that read, 'Rates: $10/hr., $20/hr. if you watch, $30/hr. if you help.' Nevertheless, I persevered and proudly delivered a final design that was both clever and simple (which is what an engineer always strives for). My solution required a few innovations, however, and the owner questioned the architect at length as to why a mechanical engineer was needed in the first place. The architect, who is a good client and a friend of mine, came to my defense.

"The solution to this kind of nightmare is to clarify lines of authority at the beginning of a project. The owner should be told that all communication must be made via the architect. Of course, this is best done diplomatically, because the owner is my client's client."

below a reasonable level and try to convince the consultant that the job has other redeeming qualities. You should be wary of tactics that usurp your right to determine price, pace, and methodology.

In dealing with clients' expectations, learn to weigh the reasonableness of your clients' requests. As your experience broadens, you will develop a better sense of give-and-take and feel more comfortable in holding your ground in the face of unreasonable demands.

ON "HAPPY" SITUATIONS IN CONSULTING

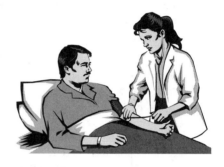

In some respects, the consultant's situation is similar to that of a doctor. A doctor does not see healthy people as a rule. Likewise, your clients will call when something has gone wrong and a quick solution must be found. The situation may have been neglected for a long time, and you may blush at your client's lack of foresight or understanding. But, just like the doctor, you must take the situation as you find it, even if your only service is to pronounce the patient dead.

There are two ways of looking at the client's problems. On the one hand, it is annoying to be exposed to problems that could have been avoided by foresight, intelligent planning, or proper attention to detail in the first place. On the other hand, if all companies had their acts completely together, the need for consultants would be considerably lower. Never forget: *The client's problems are the consultant's blessings.*

Once the consultant has worked on a problem for a while, he may begin to see how things got to be the way they are. This is not really the consultant's prime concern. His aim is to produce useful results within, and sometimes despite, a negative environment. That is, the consultant should not expect the client's encouragement to endure a political (or similarly negative) work environment. Unlike the direct employee, the consultant must be 100 percent self-motivated.

As we saw in the Michelangelo story, intelligence and resourcefulness are not always the critical factors in the ultimate success of a job. The client's ineptitude, mismanagement, or internal politics may negate the consultant's efforts. For example, a consultant may be called in to provide a third opinion to settle a dispute within a client company. Much of the input information may be biased by the individuals involved and their situation within the company. The consultant's results and recommendations will therefore have a bias that is outside of his control. The consultant can protect himself in this situation somewhat by stating all input information, sources, and assumptions at the beginning of his effort. The consultant could even organize an interface meeting to gain consensus on the fundamental assumptions and parameters. (This is easier to do in theory than in practice, because individuals who don't want to cooperate will find a way to torpedo your efforts!)

There is no surefire way to avoid occasional situations where you will offer biased recommendations or faulty conclusions. Not all your projects can be outstanding successes. Doctors are more familiar with this difficulty than engineers. Some patients may feign sickness or may give the doctor a false or inaccurate account of the symptoms. The doctor does his best during the appointment, makes his recommendations, sends in his bill, and does not worry about losing time or being unable to solve the real problem. It should be emphasized that the doctor cares about doing a professional job. This does not mean, however, that he must embrace his patient's foolishness, ineptitude, guile, or ignorance.

> **THE CLIENT'S PROBLEMS ARE THE CONSULTANT'S BLESSINGS.**

The key issue here is the consultant's *expectations.* EXPECT to work in less-than-ideal situations. EXPECT that you will have to compromise your effectiveness (but not your ethics!) to interface with your customer's pace and management style. Learn to develop communication and diplomacy skills so that you can explore solutions and approaches of mutual benefit to you and your client.

ON FIXING MISTAKES MADE BY THE CONSULTANT

Everyone makes mistakes. Sooner or later, you will, too. Every consultant eventually makes technical mistakes that seriously affect client relations and threaten chances of future work. Suppose your client calls and says, "Bill Jones here was reading your report. He has found a serious error in your calculations." How do you respond?

I prefer the good-faith approach in this situation. Speak to the individual involved and ask for the details. Find out which data, calculations, designs, or assumptions are incorrect. Determine the scope and severity of the problem. Is it a simple typographical misprint? Is it an error in assumptions? Have you used the wrong equations? Fifty percent of the time, the error turns out to be attributable to the client. If this is the case, simply explain your position. Offer to accommodate the revision (at the client's expense).

In situations where the error is yours, how you proceed depends on whether the contract is fixed-price or time-and-materials. Many fixed-price contracts have clauses that effectively bind you to fixing the work until it is "right." I can recall a twenty-person consulting company that had to redo a large fixed-price job. The error was a whopper that occurred early in the project and propagated into all subsequent results. The consulting company reworked the problem at its own expense.

For time-and-materials work, there is usually no penalty for making mistakes. That is, if the client discovers an error, you are not *legally* constrained to make it "right." Most clients will not insist that you fix the mistake *for free*. However, the politics of the situation often demands that you make concessions if you want to keep your credibility and the client's business. In such cases, I offer to split the difference. If it takes me ten hours to fix the problem, I charge for only five.

DEALING WITH CLIENT EMPLOYMENT OFFERS

You've been doing exceptionally good work for a client for three months. One day the client's vice president of engineering invites you into his office and compliments you on your efforts and talents. He explains that he could use your talents on a more frequent basis. And then he leans back, smiles, and says, "Seriously, what would it take to get you here as a full-time employee?"

Before you respond, you should think about the general politics of the situation. Most clients have *no* idea of how much successful consultants earn and how much independence they enjoy. They may be prepared to offer you so much less than you are currently earning that it will only anger you and embarrass them.

Regardless of your inclination toward direct employment with that particular company, thank the person for the compliment and tell him that you have a commitment to remaining a consultant. Playing hard to get will not detract from your chances of employment with that firm if they are sincere about offering you an attractive position. However, it *will* screen out an effort by the company to pick you up at a bargain price. Some clients feel they are doing you a favor to offer you a job on their team. After all, they're great, and they can't understand anyone not wanting to be part of their team. They think they are doing you a favor because you must be starving as a consultant. (It is also worth mentioning that some engineering firms give long-term contracts to consultants with the unspoken intention of converting them to employees.)

Tip

Another possible reaction to client employment offers is to suggest that the client give you a *retainer.* That is, he agrees to pay you for a guaranteed number of hours per year, which you, in turn, reserve for him. Such an arrangement is generally to the consultant's advantage.

If the client is sincere about an attractive position, he will persist. That is, he will explore bidding high enough to interest you. From a negotiating viewpoint, it is better to let the other person describe the kind of position he has in mind, in terms of responsibilities, title, and salary. This way, you are not put in the awkward position of asking for a salary significantly higher than that of your prospective supervisor.

Accept an offer only if it is truly exceptional and you value the particular experience and exposure it gives more than the opportunities you have in consulting. Do not dismiss the offer quickly or callously, as you may hurt your chances of further consulting work with this client.

To minimize client employment offers, do not perform your work at your client's facilities on a five-day-per-week basis for more than a month at a stretch. The consultant image will wear thin, and you will appear to be one of your client's troops. Performing some, if not most, of the work at your own place sets an appropriate distance between you and your client. It also allows creation of your own work space, which is an important factor in your sense of professional identity and self-worth.

ON FORECASTING THE END OF A PROJECT

It is essential to develop a sense for the *timing* of your interaction on a project. Without this, you will find yourself waiting for a project to begin with no billable work in hand or having a project extended while you have already made a commitment to be somewhere else on a new project.

Starting the project and performing the work on schedule is just one part of your success. Another significant part is determined by the timing of your departure. It seems that engineering and software projects always take longer than anticipated, often by factors of three to five. You won't know exactly when the project will be over, yet you have an obligation to finish what you have started. As one contract winds down, you should start to line up others, each with anticipated start dates. How does it all mesh together? It doesn't!

Don't despair. I have a corollary of Newton's First Law that applies to engineering projects:

HARVEY'S COROLLARY

Engineering projects are *inertial!*
They are late to start and late to finish.

Knowledge of this corollary allows you to become better at meshing project endings and beginnings. Thus, when starting a new contract, if you and the client agree a month in advance that work will commence on a certain date, it is far more likely that it will actually begin a week or two later than the specified date. Likewise, if you have agreed that you will be finished on a certain date, it is likely that you will actually finish weeks (and sometimes months!) later.

There are many reasons for the inertia. On the beginning side, it takes time for purchase orders to be approved, for the prerequisite project planning and management jockeying to occur, and for materials to arrive. On the finishing side, the project may simply be more involved than expected, or problems can occur with manufacturing, testing, and debugging. Cash flow problems can delay work, vendors can be late in supplying needed parts or data, and management or personnel changes can throw off the schedule.

Knowledge of the inertia corollary allows you to plan for what will actually happen. If you get caught in a bind where two clients absolutely can't manage without you at the same time, try to negotiate splitting your time between the two. Neither client will like this very much, but you have to define your own interests in this case. Simply letting a new project go because you are held up on a nearly completed one is a poor way to expand your business.

CHAPTER 12

Financing a New Consulting Practice

A REALISTIC LOOK AT MONEY

To those unfamiliar with the business aspects of consulting, financing the practice may seem as simple as writing a monthly paycheck. It doesn't take very long to realize, however, that financial matters require considerably more attention. Financial resources for the beginning consultant are usually limited and precious. The transition from regularly paid employee to self-employed businessperson can be frightening if major demands on these limited resources are not anticipated. In establishing a practice, financial commitments must be viewed as part of the overall picture, so that priorities and schedules can be made. The beginning consultant must consider *what* expenditures are necessary, from *where* she intends to obtain the funds, and *when* she should make her purchases.

THE NECESSITY OF FINANCIAL PLANNING

A preliminary business plan is a necessity if you are contemplating consulting as a full-time activity. This plan describes the tasks and resources involved in marketing, setting up an office, obtaining equipment, performing the work, and managing the business. The plan should also consider contingency action for business levels higher or lower than initially projected.

Many of the tasks in starting out are interrelated, and the planning process helps clarify the sequence in which decisions should be made. For example, before a brochure and stationery can be ordered, the company name, address, and phone number must be established.

WHAT'S IN A BUSINESS PLAN?

A *business plan* is a document that explains how you intend to package your services, penetrate your market, manage essential business functions, and

assure satisfactory financial performance. Making a business plan
is a good way to check that your
consulting service is well conceived on a *business* (as con-
trasted to a *technical)* basis. Your
plan may alert you to the need to
refine your marketing effort, raise
more start-up capital, recruit help
to perform the work, or pretest the
profit potential.

Have a trusted friend or adviser review your plan. Outside perspectives
can identify essentials that you overlooked and dramatically improve the
chances of your business success. Some independent consultants feel that
they do not have to write a business plan as long as they are using their own
money. I disagree with this attitude. Without planning, you are leaving
important aspects of your business operation *to chance.* Would you buy stock
in someone else's company if you knew that they had no business plan? Of
course not. Why treat your own money any differently?

A simplified business plan outline for consultants is presented in Table 5.
You are invited to answer the questions in Table 5 for yourself, bearing in
mind that your consulting practice may involve additional aspects not discussed. The plan is not intended to be a day-to-day guide to managing the
business, but rather a *projection* of the cash flows and the potential profitability of the activity. It is also fair to mention that profitability is not the
only factor in becoming a consultant. Many senior executives become consultants near the end of their careers so they can visit the "old ranch" on a
legitimate and useful basis. Profitability and growth could not be further from
their minds!

Tip

If you have difficulty finding an appropriate friend or adviser to review
your business plan, consider using SCORE (Service Corps of Retired
Executives), which is operated by the Small Business Administration.
On a face-to-face basis, you can confer with a retired executive who
will review your business plan at no charge. SBA offices are located in
most large cities across the country.

TABLE 5. Business Plan Outline for Beginning Consultants

1. **Define your business objectives.** Define the consulting service you are offering. Of what does it consist? (For example, design, testing, computer analysis, evaluation of personnel, supplying a product.) Of what value is it to your customers? How does the customer buy your service? By the hour, by the job, by units of product supplied or serviced? What income do you desire from the business? What are your long-range objectives for this business? How do you visualize this business ten years from now?

2. **Specify your market.** How large is the current market for the kind of service you offer? Who are your potential customers? Where are they located? How do you know they can use your services? (A good way to find the answers: Look at competitors who are already doing what you want to do.) Do long-term trends show this market area to be growing? Is your specialty on the cutting edge of technology? What market share do you hope to achieve? Have you established a price structure for your service?

3. **Describe the competition.** Make a list of businesses offering similar services. For each one, describe the niches they have cornered and the competitive advantages they maintain over the others.

4. **Outline your marketing strategies.** How will you reach customers and persuade them to use your service? How does your service offer something distinct to give you a competitive edge? How do you intend to build your credentials and gain more professional exposure to potential customers? Have you made a marketing plan (see Chapter 6) that details the steps you will take?

5. **Sketch your daily business operation.** How will you get the work done? What kinds of help will you require? How much will you use outside services? What will they cost? How will you set up your office and equipment? Will you work out of your house or rent commercial space? What equipment will you require? Will you buy or lease it? What kinds of insurance will you be required to carry? How much will it cost?

6. **Describe how your business will be managed.** Who will be responsible for making important decisions? Who will write the checks and disburse payroll? Who will administer taxes? Who will keep records? Who will perform the marketing and advertising functions? Who will review contracts?

7. **Define how your business will operate financially.** What is its cash flow and profit potential? (At the very minimum, you will need a sales forecast, a cash flow analysis, and a pro forma income statement. The sales forecast should list prospective customers and the volume of business expected from each. Be realistic!) For the first year of operation, on a month by month basis, estimate your sales volume, cost of sales, gross margin, operating expenses, net pretax income, taxes, and net income after taxes. How much working capital will you have available?

8. **Determine your form of business ownership.** (See Chapter 13 for details.) What form of business ownership will your consulting practice assume? (A sole proprietorship, a partnership, or a corporation?) Who is supplying the start-up capital, and what obligations will you incur as a result? Who is responsible for the company's debts if the business fails?

9. **Establish a contingency plan.** What happens to your projections if the basic business parameters change? Develop alternate plans to accommodate miscalculations and unforeseen events.

One of the greatest uncertainties in the plan is the number of billable hours you will achieve in a year. Here you have to make a distinction between "plan" and "goal" figures. I recommend *planning* for 1,200 billable hours, and *shooting* for 2,000, in your first year of operation. Which is to say:

Make an extra effort to stay busy at the beginning! After the first few years of operation, a goal of 1,200 hours per year may prove more satisfying.

Plan for flexibility in your schedule, as it's often difficult to predict when jobs will overlap or gap. Working more than forty hours per week for an extended period is taxing; it is a sure prescription for burnout.

The business plan is written with a particular audience in mind and structured accordingly. If you are seeking venture capital or a loan from the bank, emphasize the cash flows and potential profitability. I do not encourage you, however, to start your consulting practice with bank loans. It has often been said that a banker is someone who won't lend you money until you can demonstrate that you don't need it.

For most consultants starting out, the intended audience of the business plan consists of self, accountant, and trusted business advisers. If you are starting a partnership, the need for a detailed plan is greater. The more partners or company founders, the greater the need for a detailed business plan that explains where the money is coming from and where it is going to.

HOW MUCH MONEY DO I NEED TO START?

It is difficult to generalize about the amount of cash required to start a consulting practice. The factors leading to an estimate include: the particular type of your consulting, your personal financial situation, your initial business level, the market demand for your specialty, and your courage and resourcefulness. Some people start out with essentially no assets except their good reputation and make it. Others start out with large savings accounts and yet fail after a year. This is not to discredit the activity of planning itself, but to emphasize that planning deals with *intentions* of future actions. Actual *performance* can be very different from intentions!

Because many prospective consultants are unfamiliar with the start-up costs in the business plan, I'll describe them in greater detail. Table 6 is a list of the items most frequently needed in establishing a new practice. These items are one-time occurrences and are distinguished from the routine costs of doing business described in Chapter 9.

Table 6 is a shopping list without stated prices. For planning purposes, let's assign ballpark values for each of the items. The minimum cost of setting up an office (item 1) is about $5,000 in 1998 dollars. This cost includes the outright purchase of high-quality office furniture. I do not recommend

TABLE 6. Consulting Practice Start-up Costs

1. Setting up an office
   - Rent and lease deposit for rented office, or
   - Cost to adapt room in house for home office
   - Computer, printer, fax machine, copier, and office supplies
   - Office furniture: desks, chairs, files, tables, lamps, drapes, rugs, and decorations
   - Business insurance and phone service setup
2. Marketing expenses
   - Brochures and printing
   - Time/expense of meeting new clients
   - Wardrobe improvement allowance, if required
3. Business expenses
   - Incorporation, legal, and accounting expenses
   - Equipment, software, and reference books
   - Required licenses and permits
4. Cash reserves to
   - Cover living expenses until revenues flow
   - Pay estimated taxes
   - Cover contingencies

leasing furniture for the beginning consultant. It is much better to purchase the absolute minimum at first and add to it as you go along. Further, if it is possible to borrow furniture, supplies, and equipment from your home or from friends, do not hesitate to do so. You may need the extra cash in the beginning, and you can always purchase these same items later.

Office costs are one area where frugality pays off. Always be on the lookout for ways to keeps these costs as low as possible. For example, for the first two years of consulting, I used an economy desk that I found at a tag sale for $35. My third year of consulting was very lucrative, so I was able to purchase a quality executive desk and write it off as a depreciation item on my taxes.

Marketing, as explained in Chapter 6, is an ongoing activity for the consultant. It never ends. But it does *begin* with the preparation of consulting résumés, brochures, and sales materials. It takes substantial time, energy, and expense to meet with new customers, take them out to lunch, prepare proposals, and talk with clients over the phone. The "wardrobe improvement allowance" is a *must* if your existing wardrobe is not appropriate for meeting clients in a corporate environment. (Note: Wardrobe is not a tax-deductible business expense, but it is an essential that must be planned for nevertheless.) For planning purposes, a minimum of $2,000 is recommended for item 2.

In item 3, business expenses, some priorities are higher than others. Starting a business checking account is a practical necessity for the full-time consultant. Chapter 13 discusses the pros and cons of using an accountant to set up your company books and record-keeping procedures. For now, suffice it to say that if you *do* use an accountant to set up your records and advise you on tax savings strategies, you might spend $500 to $1,000. Do-it-yourself types will have to dedicate perhaps fifty hours of research to make the same decisions. They will also need to experiment with different accounting programs to find one that addresses their particular needs in an easy-to-understand format. Either way, setting up your books requires significant money and/or time.

Finally, almost every technical specialty requires some equipment. For example, if you are consulting in pressure vessel design, you will probably want your own set of American Society of Mechanical Engineers (ASME) Boiler and Pressure Vessel Codes. You can no longer use the set your former employer provided for you. If you cannot borrow these codes from another consultant, you will have to buy a "starter" set that costs about $500 to $1,000. Of course, always try to delay equipment purchases until you have established a definite need for them. A minimum of $2,000 is suggested for planning on item 3.

Item 4 details cash reserves for salary, taxes, and contingencies until business receipts reach adequate levels. If you already have a large contract or have already started evening consulting projects, two months' cash reserve for living expenses may be adequate. If you have no contract firmly in hand and have had no previous evening consulting income, six months' cash reserve is recommended. For typical monthly living expenses of $3,500, the recommended cash reserve ranges from $7,000 to $21,000 in 1998 dollars.

Adding items 1 through 4, the total start-up budget for a full-time practice ranges from $16,000 to $30,000. These figures may seem high. However, consider the following:

- First-year gross revenue may be as high as $200,000 (per person), so start-up costs are often small in comparison.
- If your spouse or family has income or capital that you may share, access, or borrow, the cash reserve requirements may be lowered accordingly.
- Sixteen thousand dollars is less than the price of a new Chevrolet or Ford. If you have been working for ten or twenty years in a responsible position and do *not* have this amount of cash available, ask yourself why. Would consulting be a better way to generate *more* income?
- These figures represent a low-risk transition to consulting. Many brave individuals will start out with essentially no capital and be successful. The failure rate for these brave souls is also much higher.

INC. magazine offers some perspective on the amount of capital needed to start a business. The October 1992 issue cites statistics for the 500 fastest growing privately held companies in the United States. This group of successful companies averages 145 employees and annual sales of $16.8 million. Considering the group as a whole,

- 13 percent started with $1,000 or less capital
- 34 percent started with less than $10,000
- 59 percent started with less than $50,000

The sources of seed capital are also of interest:

- 55 percent used their personal savings
- 24 percent used the resources of family, friends, or partners
- 10 percent used bank loans
- 4 percent used venture capital

Many entrepreneurs in this group started their business before the age of thirty. Over the next ten years, the trends toward youth and shoestring operations will continue to grow. Young entrepreneurs have little to lose and much to gain. Further, young business owners have lots of energy to make their dreams come true.

Since the *INC.* 500, as a group, contains few consultants, these statistics are only qualitatively applicable. Consultants, in particular, will be older on the average because their trade is more experience intensive. Finally, it is significant that *INC.* surveyed only the top 500, the "winners." Through a combination of skill, hard work, contacts, and luck, they made the most of their starting capital. The ten million "also-rans" in this contest were not mentioned. It is fair to assume that they were not as effective in reaping returns on their seed capital. Surveying the winners paints a very optimistic scenario with respect to start-up capital. If you surveyed the ten million "also-rans," you might find that the failure rate for those businesses with $1,000 start-up capital was 95 percent.

The stories of undercapitalized business failures are *legion.* However, it would be presumptuous of me to advise you not to venture for lack of cash assets. My own inclination is to avoid high-risk situations because I don't enjoy failing. But this minimal-risk strategy of mine is also limiting. If you really want a goal, and the only way to accomplish it is to *risk,* first ask yourself: What is the worst that can happen if it doesn't work out? If that consequence doesn't seem intolerable, go ahead, by all means!

CASH FLOW IN A TYPICAL BUSINESS PLAN

Example 16 shows an abbreviated version of a typical consulting business plan. The format of this plan has been simplified to illustrate the basics. The key supporting calculation is given in Table 7, the twelve-month business forecast. It lists potential contracts in this one-year period and the estimated

---

Business Objective:

Earn $60,000 per year, after expenses, by heat transfer consulting in the aerospace and electronics industries. Long-term objective is to develop salable inventions for household thermal conservation.

Strategies:

- Unique marketing twist comes from having written two computer programs that greatly simplify certain specialized heat transfer calculations. No competitor has this capability at the moment.
- Have already done moonlighting for two clients.
- Have $20,000 contract, signed and delivered, with ABC Corporation for thermal design work.
- Arrange market visits to XYZ, Ajax, and MEM to further discuss their requests for proposals.
- After three years, take 25 percent time off to pursue inventions.

Present Financial Position:

- Liquid assets available for business use = $25,000
- Spouse's liquid assets, for emergency only = $10,000
- House equity = $80,000
- Summer cottage equity = $20,000
- Living expenses, detailed on separate page = $3,000/month

Projected First-Year Balance:

- Start-up costs: During first two months, spend $3,000 per month. Subsequent ten months will require $1,000 per month.
- Monthly expenses: $3,000 for living expenses, plus $1,500 monthly contribution toward estimated taxes that must be paid quarterly.
- Business income projection: Income from ABC contract starts three months after transition day, giving $5,000 per month for four months. After that, see Twelve-Month Business Forecast (Table 7).

EXAMPLE 16. Sample business plan.

TABLE 7. Twelve-Month Business Forecast

Client	Contract Value	Duration in Months	Monthly Income	Percent Chances	Expected Monthly Income	Expected Gross Income
ABC	20,000	4	5,000	100	5,000	20,000
MEM	50,000	5	10,000	70	7,000	35,000
XYZ	16,000	2	8,000	50	4,000	8,000
Ajax	18,000	3	6,000	50	3,000	9,000
			EXPECTED GROSS INCOME FOR 12 MONTHS			72,000

probability of landing each one. The *expected value* of each contract (the actual value multiplied by the probability of getting the contract) is used to forecast the cash flow. Because the probabilities are never known with certainty, forecasting is an art rather than a science. Nevertheless, with some practice, your ability to make such predictions will improve markedly.

Figure 6 shows the all-important cash flow projection. Examination of this figure will convince you that determination of yearly income versus expenses is too gross a simplification on which to base business decisions.

You have to look at a month-by-month cash flow to see if you can keep your head above water, especially in the crucial first months. For example, in Figure 6, at the end of month 7, the accumulated income is $20,000, but the outflow is $39,500. This means that you need cash reserves (or borrowing power) to make up the difference of $19,500.

JOBSHOPPING AS A MEANS TO AVOID CASH FLOW CRUNCH

I know many people who used jobshopping as a means of eliminating the cash flow crunch in starting their consulting practice. In most cases, the compensation these people received as jobshoppers was significantly higher than the salary and benefits they had earned as direct employees. Also, in leaving direct employment for the jobshop, they were virtually guaranteed a first contract lasting four to twelve months. The security inherent in a first contract of this size provided a mechanism to overcome the cash flow crunch.

In my opinion, although jobshopping is *contract* employment, as opposed to direct employment, it's still not consulting. In jobshopping, the shop usually acts as your agent with respect to marketing, invoicing, insurance, payroll, and government security clearances. For these essential activities, the

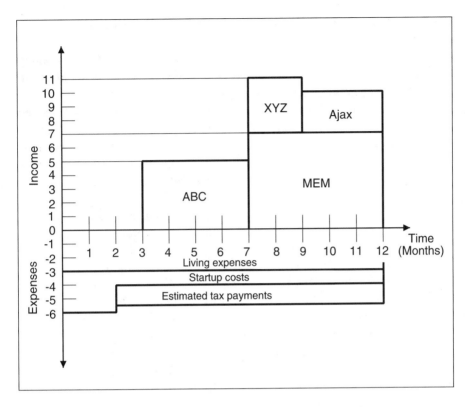

FIGURE 6. Cash flow projection.

shop claims 25 to 35 percent of the billing rate. Moreover, as a jobshopper, you will probably not be supplying expertise in the same sense as the consultant. It's more like filling in for a direct employee who was laid off the year before.

I have no objections to jobshopping for short periods of time. In the past few years, the contract houses that offer these positions have experienced an explosive growth. As corporations downsize more and more, the demand for temporary technical services skyrockets. You might consider jobshopping in your contingency plan in the event your marketing efforts stall getting off the ground. But be aware that jobshopping is not a high-status role in the technology game. Most companies that use shoppers watch them very carefully, like concubines in a sultan's harem. As a jobshopper, you may be forced into punching a time clock and other relatively unprofessional practices. Further, extensive jobshopping may label you as a "contract worker" rather than as a "consultant."

TAKE THIS QUIZ TO EVALUATE YOUR BUSINESS PLAN

You should spend a full day constructing your business plan. It is absolutely essential that your plan be in writing. It should be about five pages in length and include all important supporting calculations and subplans.

For the big-time entrepreneur, a business plan is the traditional "baited hook" to attract venture capital. Although your plan will be similar, I do not mean to imply that you should take it to your local banker, seeking a business loan. For prospective consultants, the business plan serves to clarify business and financial aspects otherwise overlooked. It will make you more comfortable about starting your own business.

The statement of objectives in your business plan is a key factor. The objectives must reflect *your* personal aspirations, whatever they are. (In a certain sense, your objectives are "you.") Your business objective may be creating lots of income. It may be to create a partnership where you can work together with a crew of trusted associates. It may be to create an ongoing business that is to be sold in five years for its asset value. Or it may be to concentrate in a specialized technical area for personal, rather than financial, reward. The important thing to realize is that the objectives you choose will affect the basic strategies you adopt in your plan.

When your plan is finished, put it aside for a few days, and then critique it using Table 8. Ask yourself whatever additional questions are appropriate to your particular situation.

FIRST THINGS FIRST: YOUR FIRST CONTRACT

One of the most important factors in starting out is having a sizable contract *in hand.* By this I mean that your first contract should be a legal certainty (not a vague or verbal promise of work) before you venture into full-time consulting. Everything else is secondary to the fact of actually doing business. Bankers and financial experts are quick to point out the difference between a planned activity and an ongoing business and assign near-zero value or credibility to the former. People look at your business differently when they hear the cash register jingling!

TAX ANGLES

Consultants in private practice enjoy considerable tax benefits. As your own business entity, you are entitled to deduct all legitimate business expenses from your gross income in computing your taxable income. The Internal Revenue

TABLE 8. Self-Evaluation of Business Plan

1. Is your business objective clear and realistic?
2. Is the projected cash flow sufficient to keep the business afloat?
3. Do you have a marketing strategy and plan as outlined in Chapter 6?
4. Do you have adequate financial resources/credit for start-up?
5. Is it better to go ahead now? Or should you wait until some major aspect of your life is different?
6. Are there "mechanisms" that would allow your plan to be more favorable?
   - Bank loan
   - Finding a partner
   - Hiring subcontractors to handle aspects that you can't do yourself
   - Using sales reps
   - Finding new applications for your skills
   - Jobshopping for first year
   - Marrying a wealthy person
   - Remortgaging your home
7. Does your plan have enough flexibility and margin for error to accommodate unforeseen events?
8. Do you really want to be in business for yourself? Can you visualize yourself as being happy running your own company?

Service (IRS) has specific rules for taking deductions, given in IRS Publication 535, *Business Expenses*. Deductions do not mean that your expenses are "free"! Think of deductions as a means of buying your business necessities at a discount whose exact magnitude is determined by your tax bracket.

Nevertheless, there are many direct benefits. You can now deduct expenses for:

- Professional books and publications
- Memberships in technical societies
- Office supplies
- Shipping, postage, and delivery
- Travel
- Attending professional meetings
- Business insurance
- Office costs (if you rent commercial space)

You are also entitled to *depreciate* major capital assets used in, or purchased for, your company's operation. This often includes computers, lab equipment, furniture, and vehicles used for business. To report depreciation,

Tip

I have found *The Ernst & Young Tax Guide* (see the Suggested Reading List) very helpful in learning about taxes. The chapter on taxes for the self-employed provides a clear and succinct introduction to preparing your Schedule C. Although *The Ernst & Young Tax Guide* is not aimed at consulting in particular, it offers insight into many strategies that can result in substantial tax savings. It tells you, by way of citing recent tax court cases, exactly how far you can stretch the interpretation of the rules. You can be sure that Publications 535 and 334, which were written by the IRS, do not elaborate on any such measures!

you must submit IRS Form 4562 along with your Schedule C. Publication 535 explains which items are treated as deductible expenses and which are treated as depreciable assets.

One of the major benefits of self-employment is the ability to set up your own retirement fund. Two options are currently available, the Keogh plan and the SEP (Simplified Employee Pension) plan.

The Keogh plan allows you to allocate a substantial portion of your income (up to a $30,000 limit), *before taxes,* to a retirement fund. The Keogh plan is superior to the retirement plans offered by most large corporations to their employees, both in the dollar amounts that can be contributed and in the flexibility of investment vehicle. The rules governing the Keogh plan have changed a number of times in the past few years, but the general trend has been to make the plan even more attractive to self-employed professionals. Since the Keogh plan will probably evolve further in the future, I won't elaborate on the details. The Keogh plan is offered through banks and brokerage houses. They can supply you with up-to-date information and application forms.

The SEP (Simplified Employee Pension) plan also allows you to set aside up to $30,000 per year. With an SEP, however, you must also provide benefits for all employees over twenty-one years of age who have worked for you three of the last five years, including part-time and occasional employees.

The Keogh plan has different rules than the SEP and involves more paperwork. Keogh plans must include employees over twenty-one who have worked for you two or more years, but only if they work one thousand hours per year. Thus, the Keogh plan is advantageous to consultants who hire part-time help only on occasion and who want to contain benefit costs. For more details, see IRS Publication 560, *Retirement Plans for the Self-Employed,* which can be obtained free from the IRS.

WHAT TO DO WITH THE MONEY THAT COMES IN

One of the greatest delights in consulting is watching how the money piles up when your business is humming along. What you do with this money is, of course, your business, but some concerns are common to all consultants. Most corporations have an entire finance department to determine the annual budget, so that the board can decide

- whether to purchase or lease a new facility;
- how much to spend on new product development;
- how many new staff members to hire;
- how large to make the annual bonus.

Consultants have similar needs. By creating a budget, you will be able to place many of your financial decisions in proper context and eliminate confusion about your financial priorities. What follows is a brief synopsis that you should modify for your own purposes and situation.

Table 9 gives an overview budget for a typical full-time consultant in individual practice. The salary has been separated into take-home pay and taxes to emphasize that you must set aside sufficient funds to make your quarterly estimated tax payments. One of the biggest mistakes beginning consultants make is forgetting about these tax payments or underestimating the amounts involved. Once you're on your own, the IRS and your state tax authorities will not remind you that these taxes are due. If you forget or underpay significantly, they will not call or send a letter, asking for the correct amount. But they *will* penalize you heavily when you get around to filing your return.

Another item called out separately in Table 9 is the amount for your retirement fund. One of the great advantages of going into business for yourself is that you can create a retirement fund that is substantially more generous than

<u>Warning</u>

Excuses such as "I didn't know that I was supposed to do this" don't work with the tax authorities. Ignorance of the law does not relieve you from your obligation to pay the taxes, penalties, and interest that you owe.

TABLE 9. Budget for Typical Full-Time Consultant

Budget Item	Percent of Total	
Salary	55	
TAKE-HOME PAY		(35)
TAXES (IRS & STATE)		(20)
Benefits	20	
INSURANCE, HOLIDAYS, ETC.		(10)
RETIREMENT FUND		(10)
Overhead (assumes home office)	10	
Marketing	10	
r & d	5	

those available to you as a direct employee. However, unless you budget for your retirement, it tends to go neglected. In the press of daily business, other budgetary priorities often lay claim to these funds. While it is understandable to skip your retirement fund contribution in the critical first year of operation (or in any other year where your revenues are low), don't make a habit of this. Be sure to allocate some of your cash flow to your retirement fund on a regular basis.

Another common mistake made by beginning consultants is forgetting to allocate enough money to marketing and R & D. The 10 percent figure for marketing may seem high to some readers, but I assure you that big-time consulting companies spend a significantly higher percentage in their operations. The guerrilla marketing methods presented in Chapter 6 may reduce some of your marketing costs relative to the big guys, but there are no quick or cost-free ways to build and maintain client relationships for the long term. Marketing costs are very real, even though your clients may deny it. (Clients not only don't appreciate this fact, they don't even want to *hear* about such

Tip

In your benefits package, include a 2 percent incentive bonus for yourself. Be sure to reward yourself after you have finished a long or challenging assignment. In general, frugality is the consultant's best friend, but now is the time to give yourself that little luxury. You deserve it!

things from you. This is one reason you should never try to explain your rate in terms of *your* costs, but in terms of the *value* that you offer the client relative to other possible vendors.)

The answer is to budget an amount for marketing. Create a spending plan for when and how you will put these funds to good use. Think of these expenditures as *paying* for the marketing information and exposure that your business needs. You should be paying to

- find out what the competition is doing;
- meet new clients, discover their unsatisfied needs, and learn their lingo;
- print new brochures, create marketing demos, and develop credibility builders;
- get yourself into the limelight, where potential clients can see you in action and remember you as an expert in your specialty.

Finding ways to achieve these marketing functions at low cost is wonderful, but don't be afraid to spend a little to make a lot. If you ask around, you will discover that most consultants consider their marketing budget to be a very worthwhile investment.

Finally, budget about 5 percent for your R & D efforts. Just like marketing, this is not the place to skimp. Spending money on developing new skills and services is investing in yourself and your professional future. This is especially important in today's fast-moving markets. Technology is changing so rapidly that you cannot coast on your present set of skills and services for very long without running dry. After setting an overall R & D budget figure, create a detailed R & D spending plan that spells out the seminars you should attend; the equipment, books, journals, and computer programs you should buy; and the development efforts you should pursue. Your annual spending plan promotes effective decision making about the technical areas you can afford to get into, considering your present situation and resources.

CHAPTER 13

Setting Up Your Business

The essence of starting a business is simply getting and doing the work, in contrast to planning or talking about it. Nevertheless, when it comes to the business aspects of your operation, it's advisable to make a few basic decisions before you leap into action.

CHOOSE YOUR FORM OF BUSINESS OWNERSHIP

The form of ownership that you choose for your consulting practice is one of the most important decisions that you will make. It directly affects your tax status, accounting procedures, insurance needs, and the way you obtain help or hire employees. The subject is complicated by the fact that the forms of business are governed by states (not the federal government) and vary from state to state. What follow are my generalizations of the decision trade-offs for the various forms of ownership. I strongly advise you to find out the rules and options that apply to your state before you make this decision.

Sole proprietorship is the most common form of business ownership in the country. According to the National Bureau of Labor, there are over fifteen million sole proprietorships in the country, most of which employ fewer than five employees. Sole proprietorship is also the easiest and most hassle-free form of business ownership. There are very few forms to fill out, especially if you have no employees. All you must do is file a federal and state Schedule C with your personal tax Form 1040.

Of all the forms of business ownership, sole proprietorship offers consultants the most control over the day-to-day operation of their business. As proprietor, you have complete control over all decisions such as company name, where to locate your office, how to allocate incoming revenues, and which clients to pursue. As soon as you have partners or stockholders, your voice in these matters is less influential. In addition, sole proprietorships are subject to fewer governmental regulations and reporting requirements.

Tip

Getting a bank loan for starting a sole proprietorship is extremely difficult, even if you have sterling credentials and a well-conceived business plan. If you are short on cash resources, the way around this obstacle is simple: Take out a home equity loan *before* you leave your direct employment position.

The limitations of sole proprietorship become apparent when you want to expand into a large operation that has a substantial payroll or when you want to create a company with asset value to be sold at some time in the future. Sole proprietorship does not lend itself to raising capital. Most sole proprietors report that banks do not treat them like businesses, but like individuals to whom they will extend a line of credit only when a personal residence or bond is put up as collateral. Yet, when you expand your sole proprietorship by adding employees, there will come a time when a client stalls in making a payment. This can leave you in a cash-flow bind too big to be solved by dipping into your piggy bank.

Trying to sell your sole proprietorship after building up the business presents another limitation. Prospective buyers want to examine a *real* set of financial records to ascertain your company's profitability. However, the sole proprietor is at liberty to commingle personal and company funds, often making accurate determination of the company's profitability impossible. Corporate forms of business bypass this problem by mandating separation of personal and company equity. Further, corporations can more easily raise capital by selling stock.

Partnerships, as defined by the Uniform Partnership Act, are the "association of two or more persons to carry on as co-owners of a business for profit." Most partnerships are formalized by writing an *Articles of Partnership* agreement that defines how the partnership is to be operated. The major topics covered by this document are:

- Amount of equity contributed by each partner and percentage ownership
- How profits and losses are distributed among the partners
- Who makes management decisions
- Who is authorized to spend monies
- How liabilities are to be shared

- What happens when one partner dies or decides to leave
- How disputes are to be resolved
- Terms for additional partners to enter the partnership

Once the partnership agreement has been signed, it is filed with the secretary of state (there is usually no registration fee).

Unlike corporations, partnerships do not have legal entity status. When a partner dies or decides to leave, the partnership does not survive. Either it is dissolved or a new partnership agreement must be drawn up between the surviving partners. In the eyes of the law, a partnership is simply a group of *individuals* participating in a legal agreement.

A partnership has an advantage over sole proprietorship in that your partner(s) can cover for you when you are sick or overloaded. When your partner's technical specialties complement yours, you can bid on larger jobs that encompass the combined breadth of your specialties. Partnerships also have more business "mass" than sole proprietorships, making it easier to get credit for expansion from commercial lenders.

Like sole proprietorships, partnerships "pass through" their income and tax burden to the partners as individuals. Partnerships must still file an annual tax return (IRS Form 1065), but no partnership tax is levied. Form 1065 merely reports the overall income for the partnership and shows how it is distributed among the partners.

The disadvantages of a partnership stem from its "marriagelike" qualities. All partners can be held liable for the business actions taken by any one partner. Or, suppose you become disenchanted with your partner over a period of years. You want to move the business in a direction that your partner does not approve. If you can't reconcile, you may have to get "divorced" (dissolve the partnership) before you can exercise your new plans. If an existing partner leaves or a new one enters, a new agreement must be drafted. Further, just like in sole proprietorship, the partners have no limited personal liability.

The most popular form of business ownership for companies with many employees is the corporation. A *corporation* is an intangible legal entity created by an individual, or a group of individuals, under statutes of state law. The legal entity thus created has the power to buy and sell real and personal property, to enter into contracts, and to sue or be sued in its own name.

Since corporations are formed by state law, the requirements for incorporating vary from state to state. In many cases, it is possible to incorporate in a state other than the one in which you do business. For example, incorporation in the state of Delaware has been popular because the incorporation fee is minimal, and because Delaware does not levy a state tax on corporate income.

However, that does not mean that the state in which business is done will not levy a corporate income tax on such Delaware corporations! The state in which you do business may levy taxes on *all* corporations doing business in the state, regardless of the state in which they are incorporated. You should check with your attorney for the rules that apply to your situation.

One significant difference between sole proprietorship and corporations is that corporations can retain earnings without paying them out to the owners. These retained earnings can be distributed to shareholders as dividends, kept as liquid assets, or reinvested in the business. All regular corporations (C Corporations) must file yearly IRS returns and must pay taxes on retained earnings. In many states, in addition to state corporate income taxes, the corporation must also pay state unemployment taxes.

The only exception is the *S Corporation* (formerly called a Subchapter S Corporation). An S Corporation is a special form of corporation that allows its shareholders to avoid the "double taxation" that occurs with a regular (C) corporation. (Double taxation means that both the corporation *and* the employees pay income taxes.) At the moment of this writing, all states except Massachusetts, Hawaii, and Vermont allow S Corporations.

S corporations pay no corporate income tax. In an S Corporation, income is passed through directly to the shareholders on a pro rata basis. Of course, there is no such thing as a free lunch: an S Corporation must satisfy certain restrictions that significantly limit the way it operates. If you can live within these rules, it may be an excellent way to organize your business. Some of the main restrictions are:

- There can be no more than seventy-five shareholders.
- There can be only one class of stock (no special or "preferred" stock).
- The S Corporation is not permitted to own 80 percent or more of another corporation.
- At least 75 percent of the corporation's revenues must come from active business rather than outside investments and passive income.

For a small group of consultants, the S Corporation is sometimes preferable to a regular corporation. But there is a new form of business ownership that may be even more attractive. In the last ten years, most states have adopted a new form of business organization called the *Limited Liability Company* (LLC). Conceptually, it's halfway between a traditional corporation and a partnership. The Limited Liability Company offers the flexibility and tax advantages of a partnership but retains the limited liability features of a corporation.

A Limited Liability Company requires at least two owners (called members), who manage the organization. It is established by filing an *Articles of Organization* with your state's secretary of state, and by paying a registration fee that ranges from $200 to $1,000, depending on the state.

The Limited Liability Company has advantages of flexible management compared with a corporation. Like partnerships, it offers pass-through tax status to the owners. As of this writing, the LLC is being classified by the IRS as a partnership for *tax purposes*.[1] All profits are passed through to members to report on their individual returns. Therefore, like partnerships, LLCs have significantly less paperwork than corporations.

LLCs even have advantages over S Corporations, for the latter must divide profits and losses according to stock percentages, whereas an LLC can distribute profits in an arbitrary fashion.

This combination of features has allowed LLCs to become popular in a very short time. Check with your secretary of state (sometimes called Commissioner of Corporations) for more details and the required forms. You might also want to check out Anthony Mancuso's guide, *Form Your Own Limited Liability Company,* published by Nolo Press.

Table 10 summarizes the advantages and disadvantages of the various forms of business ownership from a general viewpoint. The next two sections discuss corporations and partnerships specifically from the consultant's viewpoint.

SHOULD YOU INCORPORATE?

Should you, as a beginning consultant, incorporate? It depends. The subjects of corporate management, law, and taxation are so vast that they are beyond the scope of this book (and this author). The best I can hope to do here is to give you my brief impressions on the advisability of forming a corporation for independent consulting.

In its simplest implementation, forming a corporation involves just ten steps. If you decide to incorporate, it may be worthwhile having a lawyer do the paperwork for you. Such expenditures occur only once, and, more important, your lawyer will know how to organize your corporation to allow the maximum legal and tax benefit. Incorporation costs vary with the complexity of the intended corporation and the state in which the incorporation takes

---

[1] The LLC is still an evolving story. As of this writing, Congress has not passed tax legislation establishing it as a distinct business entity. For the moment, LLCs are allowed to file partnership Form 1065 for tax purposes. In the future, the IRS may create a new set of forms and rules for the LLC.

place. Typically, the filing costs are less than $500, so the major cost is the lawyer's fee for drawing up the articles of incorporation and handling the other legal chores. For a simple one-person corporation, costs typically range from $1,000 to $3,000.

For determined do-it-yourself buffs, plenty of help is available. A variety of "Incorporate-It-Yourself" books have been on the market for many years. Recently, these books have given way to inexpensive yet sophisticated computer programs that do a much better job in guiding small businesses through the incorporation process. Simply follow the program's interactive "interview," and it produces ready-made forms for incorporation in the state of

TABLE 10. Advantages and Disadvantages of Business Forms

Form of Business	Advantages	Disadvantages
Sole Proprietorship	Minimal paperwork No start-up forms to file Owner has complete control Can mix personal and company money	Difficult to get credit for expansion Personally liable for company suits Difficult to sell business at retirement
Partnership	Low start-up cost Partners can cover for each other Taxed as individuals Can offer wider spectrum of services	Personally liable for company suits Difficult to get credit for expansion Lack control of partner's spending Disagreements between partners
Corporation (C)	Limited liability Ownership can be transferred Can even out income over tax years Greater benefits packages Easier to raise capital	Requires more record keeping More tax forms to file Filing costs to start up W/other stockholders, lack control
Corporation (S)	Limited liability Ownership can be transferred Corporation pays no tax itself Greater benefits packages	Requires more record keeping Filing costs to start up More restrictions than C Corp. W/other stockholders, lack control
Limited Liability Company (llc)	Limited liability Company pays no tax itself Ownership can be transferred	Requires more record keeping Filing costs to start up

your choice. The better programs include extensive tutorials, reference facilities, and "legalese" dictionaries to improve your understanding of the process. With such computerized aids, you may be able to eliminate the need for a lawyer's help in incorporating.

After you file for incorporation, the secretary of state approves your charter and your company now has official status as a legal entity. You are no longer John Smith Associates, but John Smith Associates, *Inc.* Congratulations! Those last three letters make a big difference. From this point forward, you are an officer of your corporation and—most likely—its employee as well. As the "boss," you are now responsible for payroll administration, paying workers' compensation, unemployment taxes, and social security tax. On the positive side, you are now eligible for many significant benefits that include extremely favorable retirement plans, health insurance benefits, and equipment leasing advantages. Your accountant will be able to guide you through the maze of laws and forms to the substantial payoff on the other side.

STEPS TO SETTING UP YOUR CORPORATION

1. Determine corporate officers and directors. At a minimum, you'll need a president and a treasurer.
2. Create company bylaws that specify shareholders' meeting schedules, responsibilities and authorities of the officers and directors, and voting procedures. Much of this can be taken from boilerplate forms that are publicly available.
3. Issue shares of stock. For a consulting business, issue a nominal number of shares—say 10,000—and assign a par value of $1 per share.
4. Establish a corporate banking account to handle financial obligations of the corporate entity.
5. Set a fiscal year for accounting and tax purposes.
6. Get a corporate tax I.D. number from the IRS.
7. Choose a starting date for the corporation's "birth."
8. Select a company name. Check with the state's Division of Corporations to assure that that particular name is not already being used by another corporation.
9. Determine your corporation's mailing address.
10. File for incorporation with the state government. Some of the documents may need to be notarized. Note that requirements and fees for filing vary from state to state.

In years past, the corporate form of business held distinct advantages over sole proprietorship. Key officers of the corporation enjoyed tax-free health and insurance benefits, retirement funds with much higher limits than corresponding Keogh limits for sole proprietors, and limited personal liability in the event of legal action against the corporation. Many of these advantages are rapidly disappearing. The IRS is currently considering doing away with tax-free health and insurance benefits for corporations. The Keogh plan limits for sole proprietors have been revised dramatically upward so as to make them comparable with the maximums allowed under corporate plans. And, in recent years, the limited legal liability of the officers of corporations has evaporated. Now, when a large corporation contracts with a small corporation, the fine print in the contract often stipulates that the small corporation, *and its officers as individuals,* are legally addressable. Large corporations have been sufficiently burned by small-time operators who use the corporate veil to "take the goods and run."

Business gurus have different opinions about how valuable limited personal liability is to sole proprietors. On the one hand, it seems like a good idea. But you may discover that a good insurance policy is all you need to protect yourself from *most* of the common liability traps. Moreover, operating a corporation or limited liability company is no substitute for maintaining good insurance. Just when you think that your limited liability will protect you in a potentially risky contract, you may discover that the client has found ways to pierce your corporate veil. The client's contract may specify that you are *personally* liable for shoddy workmanship or defaults on your corporation's part. The purchasing department may adamantly insist that this clause is nonnegotiable—accept it or no deal. To cover yourself, you will have to buy the (more expensive) corporate version of the liability policy you would carry as a sole proprietor.

What, then, are good reasons for a consultant to incorporate? If your gross earnings are under $200,000 per year and you have no employees, there is no reason to incorporate. The advantages do not outweigh the extra taxes and headaches. In this case, the sole proprietorship form of business is better.

There is one common situation, however, in which it is eminently reasonable to incorporate. If one of your goals in consulting is to create a business with asset value, which you intend to ultimately sell, then the corporation is the ideal vehicle. In such cases, you as owner would build the corporation's level of business and financial strength to a point where the corporation has stand-alone credibility. That is, the corporation is able to function as a moneymaker without your further guidance. You then *sell* the corporation, based on its demonstrated ability to earn profits. This has been one of the traditional paths to great wealth in our country.

MEET THE PROS UP CLOSE AND PERSONAL

Professor J. Edward Sunderland is a widely acknowledged expert in the field of heat transfer. With more than thirty years of experience in the field and over one hundred published papers, he has the ability to solve difficult thermal problems that would baffle less seasoned consul-  tants. For exactly this reason, industrial clients have long sought out his assistance in a variety of applications.

As Professor Sunderland (his friends call him Ed) approached retirement from the university, he wanted to devote more attention to consulting as a second career. Although he had done occasional consulting on a personal (Schedule C) basis for many years, he decided to incorporate his consulting practice. From his viewpoint, being incorporated offered three significant advantages:

- Credibility to his clients, many of whom are large organizations and who prefer to do business with corporations. For Ed, this is a major factor, because it shows customers that he's serious about his consulting work.
- Medical and retirement tax advantages, as well as the ability to write off equipment used to perform his consulting work.
- Ability to sell the corporation for a profit (or pass it on to children) at a later date if it grows in size.

In 1993 Ed formed Sunderland Engineering Incorporated as a regular corporation (C Corp.) He and his wife, Mary, split the chores of running the business: Ed takes care of the marketing and technical work. Mary handles the financial and administrative functions.

Ed advises new consultants that finding a good lawyer is essential to reaping the rewards of incorporation. These are not academic issues, he says, because a properly structured corporate benefits plan can save thousands per year in health premiums as well as health charges that are not covered by insurance (so-called unreimbursable charges). In seeking out a lawyer, Ed found that many small business attorneys are not well informed on the intricacies of benefits programs. General experience in corporate law does not necessarily indicate the ability to set up a small corporation to take best advantage of the tax benefits. Ed persisted in his search and finally found a knowledgeable corporate lawyer who specialized in tax law to set up his corporation.

ON FORMING PARTNERSHIPS

When I was contemplating going into consulting, a number of acquaintances were also considering consulting as a career option. Most of them were dissatisfied with their employment situation. They wanted to form a new team, a company that would be all goodness and light. We had beer sessions in which we complained about all the nasty things our employers did and how it would be great to be in business together.

Much of the motivation to talk about starting a group was that we all lacked the business knowledge and personal initiative to just *do* it. It felt safer and more sociable to consider starting out in a group. However, not much happened from these gripe sessions. Each person had his own version of what the ideal environment would be like. What we had in common were only *negatives:* a fear of going it alone and the sheepish instinct that the flock would somehow "do it right."

Please do not jump to the conclusion that I am against partnerships! A consulting partnership is great if the partners have compatible personalities, business goals, and technical specialties. (Here, "partnership" refers to both the partnership and multiperson corporate forms of doing business.)

A good partnership has the advantages of greater credibility with customers and suppliers, pooled resources, shared overhead, and mutual support. However, these advantages quickly turn to disadvantages if the group's goals and methods deviate significantly from yours. For example, your savings in shared overhead can evaporate in a group decision to purchase specialized equipment or services that help the other partners but not you.

One of the strongest reasons for going into business for yourself is to "have it your way." It is foolish to compromise this goal by entering into a poorly matched partnership. Unless you find a truly outstanding partnership situation, it is better to start alone. After you become established, you can always enter into a partnership, and you will bring the added credibility of your ongoing business to the bargaining table. Further, in starting out by yourself, you can hire subordinates or subcontract the efforts of others to increase your capabilities. This has many of the advantages of a partnership but far fewer liabilities.

Before deciding to enter into a partnership, ask yourself the following fundamental questions:

1. Why is the partnership to my advantage? Does it increase my customer base? Does it lower my cost of doing business?
2. What are the business goals of my partners? Are they compatible with mine? Will our goals be different in two years?
3. How are important decisions to be made in the partnership?
4. What legal and financial commitments are required of me? What is my financial liability if the partnership fails? If my partner leaves? If I leave?
5. Are the professional attitudes, reputation, and competence level of my partners compatible with mine?
6. Is it easy to *communicate* my thoughts about business goals to my partners? Do they value my opinions?
7. Are there areas of the partnership that others are defining as "off-limits" to me? (For example, "I'll handle the customers, you do the work.")
8. Why is the partnership to my partners' advantage? What am I bringing to the marriage?

WILL YOU NEED A LAWYER?

Legal services may be useful in writing proposals and negotiating contracts. You won't need help in your own field of expertise, so you should focus on areas in which it is not perfectly clear as to what the customer's specification, terminology, or statement of conditions implies. If you feel you are exposed to liabilities that you are unwilling to accept, first try to clarify them with the client. If you make no progress with the client in these specific issues, your next step is to consult a lawyer to understand how large and realistic your legal liability may be.

QUESTIONS TO ASK IN CHOOSING A LAWYER

- Does he specialize in small business law?
- How much experience does he have in small business law?
- Does he have court experience?
- From which law school did he graduate?
- What is his current rate and how will he bill you?
- Does he have the time and inclination to take you on as a client?
- Do *you* feel comfortable talking with him?

Warning

Do not look in the yellow pages for a lawyer! Never retain a lawyer who doesn't come recommended to you by friends who have actually used that lawyer in a professional capacity. The abundance of unscrupulous lawyers is legendary. Hiring an incompetent or unethical lawyer is worse than having no lawyer at all. If you do not know anyone who can recommend a small business lawyer to you, make it a research project for your first year in business to find one whom you can trust.

Having passed on, the lawyer found himself with the devil in a room filled with clocks. Each clock turned at a different speed and was labeled with the name of a different occupation. After examining all the clocks, the lawyer turned to the devil and said, "I have two questions. First, why does each clock move at a different speed?"

"They turn at the rate at which that occupation sins on earth," replied the devil. "What's your second question?"

"Well," said the lawyer. "I can't seem to find my occupation. Where is the lawyers' clock?"

Puzzled, the devil scanned the room. "Oh, yes!" he finally exclaimed. "We keep that clock in the workshop and use it as a fan."

Most important, remember that a lawyer's advice is exactly what it sounds like: advice. The lawyer does not assume your liabilities by merely giving you an hour's counsel. If a lawyer gives you incorrect advice, *you* are still the one who is accountable to your client. For this reason, be sure to deal only with competent lawyers who are knowledgeable in the specific field at hand.

Selecting a lawyer usually involves a free twenty-minute interview at the lawyer's office in which you discuss the lawyer's ability and inclination to handle your legal chores. During this brief session, try to fathom whether the lawyer has sufficient experience to be of help. Is the lawyer interested in working with a fledgling business such as yours? Is the lawyer giving you clear and succinct answers—or vague weasel words? Are the two of you communicating effectively? Try to gauge the personal rapport in this meeting. Avoid the obvious personality clash that can cripple a professional relationship, but don't expect the lawyer to be your buddy. Just as your clients choose you because they want a technical job done well, the focus in your relationship with your lawyer must be on professional criteria.

Like technical consultants, good lawyers are very busy and highly sought after. Don't assume that your lawyer will be available all the time. Decide whether being shuffled off to a junior associate is acceptable to you.

Lawyers are expensive! Their fees start at $75 per hour for an absolute novice and run to $300 per hour for an experienced senior partner. Typically, a lawyer with ten years of experience charges $150 per hour for standard activities such as reviewing contracts. Most lawyers charge a different rate for court appearances, typically twice their noncourt rates.

Don't retain a lawyer unless you are sure your needs will be frequent. Instead, use lawyers for specific situations as they arise. For the sake of efficiency and economy, have all necessary materials at hand when you meet.

Situations in which you should definitely use a lawyer:

- To represent you in court
- To incorporate a multiperson operation
- To initiate or respond to a lawsuit

In many cases, a more experienced consultant or a friend who works in a corporate contracts department may be able to help you with a pressing legal question, especially if it deals with explaining the terms in a consulting contract. It is usually a lot less expensive to pay for a friend's or a fellow consultant's time than to hire a lawyer.

Tip

Nolo Press, a publisher of small business legal information, has recently produced a CD-ROM that provides small business owners with a wealth of legal resources at very modest cost (about $50). The *Nolo Press Small Business Legal Pro cd-rom* includes the complete texts of four Nolo Press books: *The Legal Guide for Starting and Running a Small Business* by Fred Steingold, *Tax Savvy for Small Business* by Frederick Daily, *The Employer's Legal Handbook* by Fred Steingold, and *Everybody's Guide to Small Claims Court* by Ralph Warner. It also includes a collection of small business legal and tax forms. The browser-like interface allows you to search for specific topics. For more details, contact Nolo Press at (800) 992-6656 or http://www.nolo.com.

RECORDS AND ACCOUNTING

You may not be very concerned about keeping accurate financial records, but the Internal Revenue Service *is*. It is in your best interest to establish an accounting system suited to your individual needs. For consultants who offer services only (no products or merchandise), life is fairly easy in this respect. If you are going to consult as a full-time activity, you will need

- A business checking account
- Income and expenditure ledgers
- Files for your expense receipts and copies of your invoices

When you set up your checking account, order business-size checks (about 8 by 3.25 inches). In the eyes of your suppliers, they add credibility to your consulting business. After you set up your business checking account, pay *all* your business expenses with it. Be sure to deposit all consulting income into your business account. When you pay yourself, write a business check to yourself and deposit it into your personal account. Keep all bank statements, canceled checks, deposit slips, invoices, receipts, and payment stubs for at least three years. If you are audited by the IRS, you may need them to substantiate your position.

In twenty years of consulting, I have never been called for an IRS audit. Yet, I have received ten official "inquiries" in this time span, letters usually phrased in tough "Explain this to our satisfaction—or pay up!" language. Most of these inquiries were automatically generated by computer programs at the IRS whose sole function is to detect discrepancies.

For example, a client sent me a payment dated December 30, 1994. I received the check on January 2, 1995, and deposited it on January 3. Because I was using the cash method of accounting, I claimed this as 1995 income. The client filed their standard 1099-MISC to the IRS, claiming that payment as a 1994 payment—as far as they were concerned. The IRS computer detected the discrepancy between the client's 1099-MISC and my 1994 Schedule C. It automatically sent me a letter demanding payment of taxes on unclaimed income. I resolved this inquiry with a letter that explained the difference and a copy of my dated deposit slip. Because these automatic inquiries can be sent up to three years after filing, it is imperative to keep accurate records for this length of time.

Separate your business and personal expenditures to simplify year-end accounting. For cash purchases that are deductible as business expenses, set up a petty cash box. Reimburse yourself by filling out a petty cash slip. If that seems like too much work, ask for a receipt each time you make a cash purchase. Make sure the item (paper, pencil, etc.) is marked on the slip, as well as the date. Put the receipts into an envelope and total them at tax time to include as a miscellaneous business expense.

At a minimum, you will need some form of records to track expenses, petty cash, accounts payable, and accounts receivable. If the "cash method" of accounting is used (see the section on accounting methods in IRS Publication 583, Starting a Business and Keeping Records), keep careful records of when payments are made and payables are received. Save all payment check stubs for at least three years.

Your expense ledger consists of a list of payments made. It should have columns for payee, amount, date, and category of expense.

TABLE 11. Typical Expense Ledger

Payee	Amount	Date	Category
Microsoft	$39.95	Jan. 1, 1997	371 — Technical books
Sony	$279.99	Jan. 5, 1997	514 — Computer drive

Your income ledger is simply a list of payments made to you by your clients. Each entry should have columns for client, amount, the date you deposited the check in the bank, and your invoice number. (With the cash method of accounting, the date that counts is not when the client wrote the check, nor when you received it in the mail, but when you actually deposited it in your account.)

TABLE 12. Typical Income Ledger

Client	Amount	Date Rec'd.	My Invoice #
ABC Universal	$4,000.00	Jan. 12, 1997	645
Galactic, Inc.	$1,525.00	Jan. 25, 1997	647

You may find computer-based accounting software helpful in tracking your finances and preparing consolidated earnings, expenditure, and tax reports. Accounting programs such as QuickBooks, Peachtree Accounting, and MYOB now offer full-featured accounting capability. From a single master switchboard, they integrate accounts payable, accounts receivable, cash flow, balance sheet, invoicing, inventory, and customer contact modules. With these programs, you can easily create monthly, quarterly, and yearly reports of all financial indicators, sorted by category. In their latest incarnations, these programs have become much more user-friendly. They offer "wizards" that guide you through the maze of accounting terminology and provide extensive on-line help and tutorials. In the future, as programmers are able to give these accounting programs more intelligence, these programs may be all you need to handle your accounting chores.

I use my portable business appointment calendar for record keeping as well. This is nothing more than a small leather-bound case that holds a monthly calendar insert. There is one page for each day of the month, with sections for appointments, notes, and recording cash expenses and daily business mileage.

In Publication 334, the IRS states its rules for how your expenses must be recorded to be considered valid. It is not sufficient, for example, to record just the amount of a business dinner. You must also record the name and city of the restaurant, who accompanied you, and the purpose of the business dinner, in order for the deduction to be valid. (I don't like this rule, but I didn't make it!) Further, if the expense is over $25, you will need a receipt.

USING AN ACCOUNTANT

Find an accountant who specializes in small businesses to help you set up your books. Such services generally pay for themselves in saved taxes within the first year. Once your books have been set up, making entries is routine. If you are not inclined to do this yourself, use a bookkeeping temp to help you at regular intervals.

Your accountant will be able to help you plan your major expenditures to maximize the tax advantages. Further, he or she can help with cash flow planning. For example, near the end of the tax year, you can often gain a tax advantage by delaying an invoice until the next year, or by paying off subcontractors and major expenses early.

Although accounting is considered dreary stuff by most technical people, it is absolutely essential to maintaining awareness of your company's financial health. Solid understanding of your spending and revenue streams will help you fine-tune your business for greater profitability and efficiency. Your accounting system can help you determine which clients/services are profitable and should be expanded, as well as which clients/services are unprofitable and should be discontinued.

If you plan to have employees right from the start, an accountant is almost a necessity to get you started with payroll reporting, tax forms, and reporting procedures. If you do not use an accountant, budget at least two days of your own time for setting up a bookkeeping system. You will also need to spend at least four hours per year reviewing the annual changes to the IRS tax codes.

Here are some of the services your accountant can provide:

- Setting up books and chart of accounts
- Maintaining books on a weekly/monthly basis
- Setting up tax reporting
- Sending in quarterly estimated taxes and monthly reporting of withholding for employees
- Preparing stockholder's annual report
- Preparing personal taxes (Schedule C)
- Preparing corporate taxes
- Administering payroll
- Setting up tax and business strategy

As with lawyers, the charge rates for accountants vary greatly. Beginners with a CPA (Certified Public Accountant) start at $75 per hour. More experienced CPAs charge from $125 to $250 per hour, depending on their degree of expertise. Almost all accountants use junior associates to handle the data-entry part of bookkeeping to keep your costs to a minimum. Accounting is a very competitive business, and many small accounting firms offer excellent value packages (on a fixed-price basis) to set up books for beginning entrepreneurs.

Some accountants, especially those in large accounting houses, insist as a matter of practice that if you want their services, your books must be set up and maintained on *their* computer. This may be appropriate for a sole proprietor running a retail store that processes thousands of invoices and payments per year. Because individual consultants typically register only a few transactions per week, having your accounting firm keep the books is overkill. Your needs are too simple to warrant outside bookkeeping help. So hunt around for an accountant who is willing to work on your terms, which may involve keeping the records on your own computer.

PAYING TAXES

Most self-employed consultants do business as sole proprietors. At the end of the tax year, they include Schedule C of IRS Form 1040, in which they report their business income.

In making out Schedule C, I use the *cash* method of accounting, as it is simple and convenient. At the beginning of each calendar year, I start a new page for each category. Entries are made as the expenses are paid and as the payments are received. At the end of the year, I have most of the raw data I need to prepare my Schedule C.

TABLE 13. Simplified Schedule C

Item	Description	Entry
A	Name of proprietor	John Smith
B	Principal business	Engineering Consulting
C	Business name	John Smith Associates
F	Accounting method	Cash
7	Gross income	98,654
10	Car expenses	3,412
13	Depreciation (from Form 4562)	1,865
15	Insurance (other than health)	352
17	Legal, accounting, and subcontracted secretarial service	2,240
18	Office expenses	3,986
20	Rental (equipment)	358
23	Applicable business taxes	150
24a	Business travel	1,743
24d	50% of qualified meals and entertainment	336
25	Utilities	154
27	Other expenses:	
	Bank service charges	34
	Prof. dues and publications	675
	Prof. Eng. license fee	150
28	Total expenses (sum 8 to 27)	15,455
29	Tentative profit or loss (subtract line 28 from line 7)	83,199
30	Home office expenses (attach Form 8829)	654
31	Net profit or loss (Subtract line 30 from line 29; enter result on Form 1040, line 12)	82,545

Table 13 shows simplified Schedule C entries for a typical consultant. To fill out item 13, you will first need to finish Form 4562, in which you figure the depreciation on capital equipment purchased for your business. Expenses for a home office are reported on Form 8829 and then entered on item 30 of Schedule C. You must satisfy certain requirements to take the home office deduction: You must use the office *regularly* and *exclusively* for business purposes, and you must perform the majority of your services there. The IRS describes the details in Publication 587, *Business Use of Your Home.*[2]

---

[2] To obtain copies of this and the other IRS forms, call 1-800-TAX-FORM (1-800-829-3676). For answers to your tax questions, call the IRS information service at 1-800-829-1040. The IRS Internet site (http://www.irs.ustreas.gov/) offers forms and publications in electronic formats.

It is worth noting that the net profit amount on item 31 is *not* the same as your "equivalent salary." There are two reasons for this. First, the deduction for your Keogh or SEP retirement plan is not given on Schedule C, but is entered on line 27 of Schedule 1040 as an adjustment to income. Second, in addition to Schedule C, you must fill out Schedule SE, the self-employment tax. When you are an employee of a corporation, the employer pays a social security tax on your income, amounting to roughly 50 percent of your contribution. Now that you're self-employed, *you* pay this extra tax.

Many self-employed professionals feel that the self-employment tax is unfair. They resent it. I don't feel that way. What it says to me is: Your charge rates must be high enough to absorb this extra cost of doing business.

To reduce the likelihood of being audited, append to your Schedule C a list of clients and the total payments each made in the calendar year. (See Table 14.) Why do I suggest this? Some of your clients may send you a Form 1099-MISC at the end of the tax year, reporting the amounts they paid you during the calendar year. The IRS also gets copies of this form and will look for this income spelled out explicitly in your return. If you do not supply this list of clients on your own initiative, the IRS usually sends an inquiry letter, demanding an explanation or the taxes due on the "unreported income" (plus penalties). This short precaution avoids the hassle of having to compose a detailed letter explaining the discrepancies.

Item D of Schedule C (not shown in Table 13) asks for an employer ID number. If you operate your consulting business as a sole proprietor and have no employees, you don't need one. Leave item D blank. When you become an employer or change to a partnership or corporate form of business, you must get an Employer Identification Number from the IRS by filing form SS-4. The IRS will then start sending you quarterly and year-end payroll tax returns for you to fill out and return.

For consultants who use the partnership, corporate, or LLC form of business, the IRS has separate forms and reporting requirements. If your accountant does not carry these, ask the IRS Forms office for Publication 541, *Tax Information on Partnerships,* or Publication 542, *Tax Information on Corporations.*

TABLE 14. List of Clients Appended to Schedule C

Client	Amount
Universal Incorporated	34,765.00
Acme Research	12,350.00
Generous Motors	11,000.00
Lyben Computronic, Inc.	37,500.00
Ace Appliance Labs	1,500.00
Peripheral Engineering, Inc.	1,539.00
TOTAL	98,654.00

ESTIMATED TAXES

When you are an employee of a company, your employer is required to withhold federal and state taxes from each paycheck by an amount that approximates your tax liability. Now you're in business for yourself. You accomplish the same goal by submitting quarterly estimated taxes on Form 1040-ES. The four installments are due April 15, June 15, September 15, and the following January 15.

If your net income from consulting is very small, you may fall under the limit for being required to file estimated taxes. As a rule of thumb, you need to file estimated tax payments only if your total tax, reduced by any withholding amounts, exceeds $500.[3] For the details, see IRS Publication 505, *Tax Withholding and Estimated Taxes*. You may also be responsible for paying estimated taxes to your state government.

Suppose that in addition to consulting, you hold a part-time job as a direct employee. Your employer withholds an amount based on your earnings as a direct employee. Must you still pay estimated taxes? Yes. In this case, your estimated payments need cover only the part of your tax liability that hasn't already been covered by your direct employment withholding.

The important thing to realize about estimated taxes is that when you are paid by your clients, you are receiving *pretax* dollars. If you don't hold the proper amount for taxes in reserve, you may find yourself pressed for many thousands of dollars on the quarterly installment date. Because the penalties for late payment and underpayment are severe, it is worth monitoring your tax liability carefully.

HOW TO HIRE HELP

What should you do about secretarial help? As a direct employee, you probably shared a secretary with a few others in your group. Now it's not economically feasible to have a full-time secretary. When and if your consulting practice grows to five full-time people, you might consider such a luxury. In the meantime, for typing, filing, and other office needs, use temporary secretaries on a contract basis from a temp agency. If you must have your phone answered by a "real" person (many people are annoyed by answering machines and voice mail), consider a local phone answering service. The charges for this service are modest, and it gives your business a more personal touch.

---

[3] This rule of thumb is based on taxes. In terms of *net outside income,* a conservatively safe equivalent is: If your net outside income exceeds $1,000 per year, file estimated taxes.

For occasional technical help, I strongly recommend using the services of outside ("independent") contractors, not hiring employees. What is the difference?

When you hire an employee (part-time or otherwise), you are legally obligated to prepare an enormous amount of paperwork. By the time your business has annual revenues in the $300,000 range, you may have no choice but to hire employees. However, when you are starting out and your revenues are modest, the extra paperwork and hassle are simply not worth it. It's to your advantage to purchase help on a contract basis.

As an employer, you must keep auditable payroll records; withhold federal, state, and local taxes; prepare quarterly and year-end payroll taxes; pay employer social security, unemployment, and Medicare taxes; and carry workers' compensation insurance. (In the future, you may also be required to contribute to "portable" medical and pension plans that your employees carry for themselves.) Further, depending on the type of pension plan you have adopted, you may be required to set up a retirement fund for your employees.

When you hire an independent contractor, you are buying the service. Period. The contractor handles his own taxes, insurance, unemployment, and workers' compensation.

Now for the sticky part. The IRS has ruled that even though a person calls himself an *independent contractor,* he may actually be a *common-law employee.* This distinction has profound implications. When you hire an individual to perform a job for you (even something as small as a week's worth of typing), you should clarify your business relationship by having that person sign a subcontractor tax release. In the event you are using the services of another consultant as part of your project team, clarify this issue up front as part of your negotiations.

The subcontractor tax release (see Example 17) says that the person will handle his own taxes and that you are not liable for them. But if you continue to use that person for an extended period of time, and the person has no business other than yours, the IRS says that your relationship is one of employer-employee, regardless of what the two of you call it. Even though you may have a signed subcontractor tax release, you, as the pseudo employer, may still be liable for certain kinds of taxes. You may also be subject to a lawsuit in the event the person is injured on the job. If he doesn't carry workers' compensation for himself and you don't carry it (because you think the other person is responsible for it), *you* are the party at risk.

Table 15 shows the criteria the IRS uses to determine whether a person is an employee or an independent contractor. The IRS guidelines are explained in Publication 15, *Circular E, Employer's Tax Guide.* To satisfy independent contractor status, a majority of your answers must agree with the

Harvey Kaye & Company
42 Bent Oak Trail
Fairport, NY 14450

716-223-4502   **Consulting in Mechanical Engineering**

SUBCONTRACTOR TAX RELEASE

In consideration of services performed for Harvey Kaye & Company, I will receive gross income for which I assume total tax responsibility. This tax responsibility includes federal taxes, social security taxes, self-employment taxes, state taxes, local taxes, and any other taxes or penalties to which I may be subject.

I release Harvey Kaye & Company and its clients from any responsibility for these above-mentioned liabilities.

Name _____

Address _____

Signature _____   Date _____

EXAMPLE 17. Subcontractor tax release.

"Independent Contractor" column. You can also request an IRS opinion on employee status on IRS form SS-8. These laws are still evolving and are often successfully contested in tax court. The best people to ask for advice on this subject are established independent consultants in your network.

If you hire outside contractors, the only paperwork involved is filling out an IRS form 1099-MISC (if the person's earnings were more than $600 in that tax year). One copy of the form goes to the IRS, the other goes to the contractor no later than January 31 of the following year.

TABLE 15. Criteria to Determine Employee Status

Criterion	Employee	Independent Contractor
Does the person provide services for more than one client at a time? (By far the most important criterion.)	No	Yes
Does the person have a significant investment in facilities such as home office, tools, and equipment?	No	Yes
Can the person realize a business profit or loss as a result of his or her services?	No	Yes
Does the person make his or her services available to the general public?	No	Yes
Does the person use his or her own tools, equipment, licenses, and business cards in the course of performing work?	No	Yes
Does the client provide instructions to the person about when, where, and how the work must be executed?	Yes	No
Does the client train the person to perform services in a manner specified by the client?	Yes	No
Does the client set the person's work hours?	Yes	No
Is the service performed mostly on the client's premises?	Yes	No
Is the person paid regularly by the hour, week, or month?	Yes	No
Does the client provide the person's insurance, health, and retirement plans?	Yes	No
Does the business relationship with the client continue indefinitely (in contrast to a specific duration)?	Yes	No

The IRS is strongly motivated to classify independent contractors as employees because it increases its revenues. On the other hand, employers are motivated to classify individuals as subcontractors because they can save on employer social security taxes, unemployment taxes, and workers' compensation insurance. Subcontractors are motivated to keep their status because they can usually claim larger deductible business expenses than employees.

For independent consultants, three practical considerations result from these IRS regulations:

- Any time you hire a subcontractor, make sure he understands that he is to handle his own taxes, workers' compensation, and benefits. Formalize this understanding in a subcontractor tax release or similar document.
- Don't hire Suzy or Johnny, who lives on your street and is looking for a few weeks of work, unless she or he is a bona fide independent contractor. The few dollars you save are not worth the potential hassle.
- Never hire Suzy or Johnny to work "under the table." It's illegal. Doing so puts you at risk of criminal prosecution.

STRATEGIES FOR MAINTAINING *YOUR* INDEPENDENT CONTRACTOR STATUS

As long as we're on the subject of independent contractors, let me digress for an important aside. So the IRS won't consider *you* a common-law employee of your major client, never devote all your time to just one client in any given tax year. Exert all reasonable efforts to always have at least two (preferably five or more) paying clients every tax year. This may cause you some anxiety if you are locked into a single well-heeled client who needs you for an extended period of time. The money spigot is wide open, and you may be too greedy to take time out to market other clients while business is so good. If you neglect to create a larger client base, sooner or later you will have to do some fast talking to the IRS.

Here are some strategies for avoiding the "one client per year" trap.

1. Offer a special break on your rates for occasional small jobs that you would otherwise lose.
2. Work more than forty hours a week. If you are committed to a forty-hour-per-week effort for an extended period of time, obtain work from a different client that can be done on evenings and weekends.
3. Spend more time marketing. Any way you slice it, working full-time for a single client portends business problems. Wean the client of the idea that you will always be available forty hours per week. Then get moving with your marketing plan!
4. If all else fails, work out a reciprocal agreement with another consultant whom you trust. Say, for example, you both agree to review each other's business plan. You give the other consultant a contract and a $500 check. The other consultant does the same for you. Without cost-

ing you anything on a net basis, you now have another client. (Although this trick is not illegal, you can be sure that the IRS will frown on it. Use it with caution.)

WHAT KINDS OF INSURANCE MUST YOU CARRY?

After you start your consulting business, find a local insurance agent who handles business insurance. Insurance agents will explain the various types of coverage available, discuss how they apply to your business, and help you determine your exact needs. They can tailor a package deal for you that encompasses the various coverages you require. The kinds of insurance you need will depend on the exact nature of your work, but as a minimum you should consider

- Workers' compensation
- General liability
- Fire and theft
- Business motor vehicle use
- Business travel

Workers' compensation insurance provides benefits to employees who are injured on the job. Most states require all employers to carry this insurance. The individual consultant, as a sole proprietor, cannot be covered by this insurance; it is for your employees. If you have no employees, you may get away without it. Just remember that hiring an occasional part-time worker makes you an employer. Even if he or she works two days per month for you, you will be required to carry workers' compensation.

If you are incorporated, you are an employee of your own corporation, and workers' compensation is necessary even if you never hire anyone else. The cost of workers' compensation (for consulting companies) typically runs $300 per year and increases slightly as the number of employees increases. The premiums depend strongly on the occupation covered and on the number of claims your company filed in previous years. Workers' compensation insurance for high-risk workers, such as roofers and telephone line repair personnel, is about ten times more costly than that for low-risk occupations such as consulting. If one of your employees files a claim, expect your rates to rise significantly the next year. (This is the raison d'être for most corporate safety programs.)

General liability should not be confused with *professional liability* insurance. The latter covers the consultant's liability for incompetent or negligent performance of professional services, such as in rendering a faulty design or fabrication that causes an accident. Such insurance is very expensive and is not recommended except in special circumstances.

General liability insurance, on the other hand, is concerned with physical and property damage to both the client's and the consultant's personnel and property. It covers, for example, accidents such as your client's tripping over your office rug and breaking his ankle, or your being hit by a flying metal fragment while touring the client's fabrication facility. Such insurance is obtained at modest cost and is often required by clients in cases where you will be spending more than a few hours at their facility. You can determine if you are required to carry such insurance by looking at the "Hold Harmless" or "Indemnity against Claims" clauses in the consulting agreements you sign.

Comparison shop for your insurance policies. The insurance business is extremely competitive, and you may be able to save significantly by getting quotes from more than one qualified vendor. Ask for written quotations before you make your final selection.

SHOULD YOU HIRE A MARKETING REP?

If you offer such specialized services as

- A particular lab test
- Use or rental of your unique machine
- Use of your proprietary computer program

then you may want to use the services of a sales rep or marketing agent. Such agents represent a number of small companies and will "beat the bushes" for you. They usually charge about 10 or 15 percent commission for the business that they bring in.

I have a friend who uses a marketing rep to sell his specialty, which is a particular type of vibration analysis needed to certify nuclear power plant components. The agent extends the consultant's marketing arena in two ways: first, geographically; and second, contact-wise, which is to say that now the agent's list of contacts is a possible source of new business.

Two pieces of advice in using a marketing agent: First, do *not* use a marketing agent who also just happens to be capable of doing the work himself! If you do this, you are breeding a potential competitor. Second, do not pay commission until the work has actually been done. That is, the commission should not be for a "meeting of the minds" or even for landing a contract with an estimated value. In some cases, the contract may be terminated at less than the stated value or dropped altogether. You should pay commission on work that you have actually billed, not on work that you *hope* to bill!

YOUR COMPUTERIZED OFFICE

In today's technology-driven world, you will need a fax, a modem, and an Internet connection to communicate with your clients. Fax machines now cost less than $300 and allow you to send and receive documents electronically over the phone line. Modems enable your computer to "talk" with other computers over the phone lines. With them, you can send and receive data files, reports, and executable programs. The uses for modem applications are growing. You can now tie in to national telecommunication networks to search large databases, buy and sell merchandise, trade stocks, make airline reservations, book hotel reservations, track news items on specific subjects, and do much more.

Telecommunications hardware and software are evolving rapidly. Portable computers will soon have fully integrated mobile communications capabilities. This means that a single briefcase-size unit will house a personal computer, a fax, a modem, Internet lines, and a cellular phone. Using the modem link to centralized computers, you will have access to an enormous database of information and on-line services wherever you travel. New miniature TV cameras connected to computers will allow two-way audio-video communication over the Internet. These features are available today, but they are expensive and not well integrated into a single unit. In a few years, prices for the hardware and on-line services will drop to the point where everyone will be able to afford them.

In setting up your computerized office, the particular brand of equipment you use is not of great significance. All the major personal computer (PC) vendors now offer equipment that has improved enormously since the PC was first developed. They now sell, for a few thousand dollars, machines that can blow the doors off 1970 vintage mainframes. It is highly likely that the flood

 Warning

Using fax, modem, and Internet for instant communication of your data can sometimes be a liability instead of an asset. The client may assure you, "Don't worry, we're in a hurry. Send us your quick and dirty results." Once you have sent them the information, however, it often has a way of eventually becoming represented as your final, polished assessment. Especially in the beginning of a project, don't shoot yourself in the foot by prematurely delivering material that might demonstrate lack of judgment or attention.

of exciting new products in the computer field will continue to refine and redefine our notions of what is possible for office computing applications. At the minimum, most technical consultants should consider the following hardware for their own office:

- Personal computers, both desktop and portable
- Laser printer
- Fax and modem
- Scanner to reproduce images
- CD-ROM
- Video-conferencing cameras
- Multimedia (sound and animated video)

You will also need software for technical consulting. Beyond that required for your specialty, you should get:

Word processing to write reports and correspondence. Just as clients are positively influenced by your professional appearance, they are also impressed by neat, organized, and well-printed reports. Word-processing programs now have spelling checkers, grammar checkers, equation editors, and other conveniences that allow you to produce professional-looking reports. Many consultants tell me that clients are more inclined to believe reports that have been printed out by computer.

Spreadsheet programs to organize, manipulate, and present technical or financial data.

Database programs to organize large amounts of information. Databases allow you to query to find particular subsets of the data. Using Boolean searches, you can find the records you need quickly. For example, suppose your database contains information for four hundred sales prospects, but you want to send letters to only those individuals who have sent in the coupon *and* who live in your home state. You do a query for (Coupon = Yes *and* State = New York). The program then gives you a list of only those entries satisfying the condition.

Contact management software tracks the names, addresses, phone numbers, and other data for your prospects and clients. Since database programs can handle arbitrary sets of information, they can be set up to do contact management also. Contact managers, however, are custom-built and optimized to perform only one function—track contacts.

Many contact manager programs have a time-tracking function that lets you "clock" the length of time you worked on a particular project or the duration of phone calls with the client. For this reason, contact managers should be of particular interest to consultants. Often, consultants underestimate the total time they put in on a project by neglecting the small chunks. Fifteen minutes here and there add up. (A major source is handling telephone calls from the client.) The contact manager allows you to log phone calls, post beginning and ending times, and make electronic notes on the discussions. If these little extras add up to two hours per month, you can bill the client for the extra time, providing it isn't specifically excluded in your contract and providing you have the contact manager records (or similar time log) to substantiate your bill. Of course, if the extra time turns out to be only ten minutes per month, don't bother with it.

Graphics, charting, and presentation software will help you prepare charts, illustrations, and slides. In the last few years, graphics software has become easy to use. With minimal effort on your part, it will accept numerical data from your spreadsheets and databases and automatically generate attractive charts and graphs.

Other categories of software you may need include:

- Desktop and Internet publishing
- CAD (computer-aided design) and flowcharting
- Accounting and business planning
- Telecommunications (fax and modem)
- Internet (browsing, E-mail, group collaboration, and Web site authoring)
- Programming languages (Visual BASIC, C++, Delphi, etc.)
- Mathematical
- Project management
- Reference (dictionaries, encyclopedias, mapping and trip making, foreign language translation, phone databases, etc.)

CHAPTER 14

Office Space and Stationery Needs

WHAT'S THE BEST LOCATION FOR YOUR OFFICE?

For the full-time consultant, office location is one of the first major decisions to be made. In addressing this matter, first consider your requirements. You will need a reasonably quiet and presentable place in which to work. You will also need enough space to accommodate desks for yourself and your assistants (if you have any), file cabinets, computer equipment, phones, bookcases, stationery cabinets, copiers, lamps, whiteboards, and whatever additional equipment your specialty entails. Further, you may need a conference table with four to six chairs if you plan to hold client conferences and technical meetings at your office.

On the other hand, you do *not* need to be located in an expensive, high-traffic area. In most cases, you will be visiting your clients at their facility. Do not burden yourself with the expense of maintaining luxurious offices unless you are independently wealthy.

HOME OFFICE VS. RENTED SPACE

The two most common choices for consultant's quarters are rented commercial space and the home office. Each choice has advantages and disadvantages for the starting consultant.

The home office is the minimum cost option. You pay no rent or additional utilities such as heat, light, and telephone. Further, there are tax advantages to having a home office if it is used regularly and exclusively for business. The IRS has specific criteria to determine the applicability in each situation, given in its Publication 334, *Tax Guide for Small Business.* (I find, however, that the size of the home office deduction is relatively small.) The home office has the additional benefit of eliminating everyday commuting.[1]

With the advent of the Internet, home office workers are no longer as isolated as before. Being plugged into the 'Net is almost as good as being there. You can exchange E-mail messages with clients and send and receive documents, progress reports, and technical drawings. New telecommunication systems will allow you to hold two-way video conferencing. From your home office, user and chat groups offer a forum for meeting other professionals with similar interests.

In some situations, a home office has a number of strong disadvantages. A household with children may be too noisy and distracting to allow concentration. Your loved ones may figure that as long as you're at home, you should do the laundry during your work hours. Or they may interrupt you constantly for small favors. Also, it is usually difficult to expand a home office. Finally, the home may be unpresentable to the occasional client who visits because

- it is too shabby;
- it is too elegant;
- it has decor that sends an unprofessional image (for example, artwork depicting nude models);
- the only available space is in the second-floor bedroom area or the basement (tacky!);
- the location is not convenient for your clients (for example, no parking or too far away).

Sometimes you will not be able to determine how suitable a home office is until you actually try it. One advantage of starting your practice by moonlighting is that it allows a low-risk, but accurate, calibration of the feasibility of working at home.

Renting an office in a commercial building sounds like an ideal solution, but the cost of such surroundings can be prohibitively high for the beginning consultant. A 300-square-foot, two-room office costs from $300 to $1,200 per month, depending on location. In addition to utilities, you may have to purchase business insurance to protect your personal property on the

---

[1] *Officially,* your residence must be zoned for business use, or you must get a "variance" from the local zoning board. *Unofficially,* some ten million people run home-based businesses in direct violation of zoning regulations. Unless you do something extraordinary to call attention to your home consulting office — like erect a giant sign on your front lawn — nobody will ever bother you about this issue.

premises and to cover general liability. Finally, the lease you sign may be hard to break if you relocate, and there is usually no protection from unreasonably high rent increases after the first lease has expired.

The entire rental cost of commercial office space can be deducted from your taxes, as well as all business telephone and utility costs. (When your business volume and gross income are high, this tax advantage gives you, in effect, about a 50 percent discount.)

The trade-off between a rented office and a home office is largely dependent on your personal situation. The advantages of having a rented office are obvious. In fact, if this were not the preferred choice for working space, it would not be as popular as it is today. However, the cost of rented office space is high, and it is important to review the financial aspects in your business plan before a commitment is made. If you really must have a rented office and are concerned about the cost, explore ways to share facilities with other professionals or to rent a portion of another company's business space. A reasonable effort in searching for the right situation can have significant long-term payoff.

WELCOME TO MY OFFICE

To give you a better image of a home office, let me describe mine. My office consists of two rooms that have a total of 300 square feet of floor space. There are five windows that allow plenty of natural light and fresh air. There are areas for client reception, desk, computer 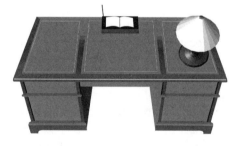 station, and office supplies. The layout is designed for maximum convenience in use. There are six large bookcases that hold frequently used technical references. I collect helpful magazine articles and technical society publications in three-ring binders according to subject. (There is also a storage area in the cellar for seldom used journals, books, and computer disks.)

In addition to the three-ring binders, I have developed a filing system for reference material as my library of technical papers has grown. Without a filing system, I had to sort through the entire pile of material to find a single item. This "sequential access" method was so exasperating that I finally put in the effort to work out an organized system.

All the furniture was selected for quality of construction and comfort in actual use. Two years after I started out, I bought a cherry executive desk that has served me well over the years. I paid a substantial sum for it but have been

Tip

Do not buy office furniture on the basis of catalog description. Pictures and words can be misleading, especially with regard to chairs, desks, and file cabinets. Before you buy, sit in that chair, try writing at that desk, and test the sliding action of those file-cabinet drawers! You won't be sorry you took the extra time to comparison shop and test the various models in the stores.

pleased at the way it has held up. I am on my third office chair, though. My first was an inexpensive model that looked great but gave me backaches. The second was of higher quality, but it started to squeak unbearably after four years. All efforts to fix the squeak failed, so I got rid of it. My present chair is a used Herman Miller that I bought six years ago at a secondhand office furniture outlet. In nearly new condition, the chair cost me $180, which is a lot better than the $1,000 or so that a new model now costs. I am thoroughly pleased with this chair; it looks great, and I can sit in it comfortably for hours without any back strain. The walls of my office are decorated with paintings and photographs that reflect my interests in art, nature, and philosophy. There is a beautiful Bokhara Oriental rug on the floor, and drapes of my choice on the windows.

One reason I have a home office is that most of my interaction with clients is at their facility. After all, I am the vendor, and in my type of consulting, it is more common for the vendor to visit the buyer.

Many large consulting companies insist that most or all work be performed at their own facility. That is, they do not want their staff consultants performing work at the client's facility. They do this for a number of reasons, mostly to retain control of overhead costs and to avoid losing employees to the client. However, for the individual consultant, insisting that all work be done at your own office is not necessarily an advantage. Being at the client's facility, at least some of the time, has definite payoffs.

A SHAGGY-DOG STORY

A few years back, I had a contract to do a heat transfer analysis for Raytheon, one of the country's largest electronics firms. One day, while I was at lunch in their cafeteria, a man joined the group at the table. He noticed my "brass rat" (the M.I.T. class ring) and held his up to mine, saying, "I see we went to the same school." We struck up a conversation, and one thing led to another. He asked about my specialization in graduate school. When I told him about my work on high-temperature gas dynamics, his eyes lit up. On the paper lunch mat, he sketched a device, told me a few facts about it, and asked me if I knew how to determine its temperature.

I asked a few questions, and then told him, in general terms, what I thought the problem was about. In a few sentences, I indicated how I would go about solving it. In doing this, I was careful to use enough buzzwords to let him know that I was familiar with the technology, and didn't let myself get bogged down in details. I relied on my technical intuition to quickly put together an approach.

It turned out that this man was the manager of a high-priority project. A week later I got a call from him. Would I write a two-page technical proposal for him, expanding on my lunchtime ideas? I filled his request and received a large contract from him just a month later!

The point is *not* that you should be an M.I.T. grad or wear a brass rat! The idea I wish to stress is that you will meet other customers at your client's facilities, if you are merely open to that possibility in normal conversation. This has happened to me so many times that as a personal rule, I never turn down a request to perform some of the work at the client's facility.

Bringing this shaggy-dog story to a close: If you are spending significant amounts of time at your clients' offices, why do you need rented space? I would consider a rented space only under the following circumstances:

- Your home is exceptionally noisy or distracting
- The available home space is too small to be effective
- You are starting out with so many employees or partners that it is awkward to have them in your home

YOUR PORTABLE OFFICE

The shaggy-dog story brings up another issue: How do you arrange for a work space when you're performing work at your client's facility for a month? In my experience, the office you will be assigned as a consultant working on a project is usually the worst one available. By this I mean worse than that of the lowest ranking direct employee, even worse than that of the lowest ranking jobshopper there for a one-year assignment.

If the client is tight on office accommodations, you may be assigned to the coat room, conference room, or even the computer closet! I am not kidding! The client will do everything in his power to make your stay as a consultant less than comfortable. You will get the worst computer, the worst chair, poor lighting, no storage space, and often no telephone.

Why? First, clients often resent having to pay so much for your help. They don't understand the calculations in Chapter 9 that show that not all of your hourly rate goes into your equivalent paycheck. They forget about the costs of your benefits, your company's G & A, marketing, and vacation costs. Most important, they neglect to account for your downtime. They know that they can turn you "on" and "off" but don't grasp that this *must* translate into a higher equivalent hourly rate!

Second, the employees at the client's office have struggled long and hard to get perks such as their expensive office furniture and their fancy computer system. These perks are part of the unspoken reward system that exists at most companies. The client will be reluctant to grant you these prized rewards at their facility just because you are visiting for a month to complete a specialized task.

Third, most clients don't have keen insight into what *your* needs are. All they know is that they committed an exorbitant sum just to get you on the project. It's like when you ask the furnace repair company to come to your house. You were upset when they said that merely showing up at your door would cost $59. You were worried that the bill might swell to gargantuan proportions at the hourly rate of $75 plus parts. And now the young, bright serviceman has the nerve to ask if he can borrow the No. 2 Phillips screwdriver from your tool rack. No way!

Furnace repair companies have solved this problem: They come complete with *all* of their own tools and parts. They don't have to ask you—the homeowner—for a single thing, except the payment at the end.

To as great an extent as possible, this is also the solution to the consultant's problem at the client's office. Unfortunately, you cannot afford to be as aloof as the furnace repair person: no van in the world is big enough to stock all the things you might need as a consultant. Therefore, learn to live with poor working conditions, noise, and lack of privacy. As a consultant, these parameters are out of your control when you are using the client's facilities.

There are many things you *can* do to improve the situation. First, develop a basic *portable office kit*. This consists of briefcase, pens, paper, ruler, stapler, paper clips, coffee mug, tissues, and whatever other staples you require. Label all these items as your personal property before bringing them to the client's office. Second, develop a portable reference library of essential books that you need. Third, develop a portable software library of essential programs you can't do without. Fourth, if you are not given a phone, bring in your portable cell phone. Fifth, bring in your portable computer. (Make sure you have a secure place to lock it while you are away from your area.)

It may not be feasible to use your portable computer at the client's office for a number of reasons. For example, you may need to be plugged into the client's network. You may require access to proprietary programs that cannot be placed on your system. Your system may be incompatible with peripherals such as printers, scanners, and bar-code readers that you may need. The client may prohibit personal or non-client-owned hardware from the premises to deter theft of company property. If this is the case, then develop a tolerance for working with less-than-optimal equipment or negotiate to perform the work at your own facility.

For me, just knowing that poor working conditions are the rule rather than the exception has proved to be very helpful. I learned a long time ago to dissociate my feelings of self-worth from the way clients handle my "room accommodations."

OPTIONS FOR OFFICE PHONES

Many consultants continue to use their personal phones for their consulting business. This is certainly the lowest cost option, but it may not be suited to your lifestyle. For example, if you have young children in the house, you may not want them picking up your business calls when you are away. When clients hear, "Hello, this is Jessica. Did you want to talk to my Daddy? He's not home now. I'm four years old and I just made a big mess in the kitchen. Mommy is going to be really mad when she sees it," they will receive an unprofessional image of your business.

Similarly, if your teenagers use the phone a lot or if you are on-line for three hours per day, your clients may not be able to get through to you. Under such circumstances, install a dedicated business phone line. Although business phone rates are higher than residential rates, they are still very affordable.

Other services you should consider are available from your local phone carrier:

- Call forwarding, by which incoming calls are forwarded to another number you designate.
- Caller ID, which displays the phone number of incoming callers. This is very useful for screening calls.
- Voice mail, which does everything that an answering machine can do and also takes messages when your line is busy.

The telecommunications field is still rapidly evolving. In the near future, you will be able to control all phone functions from a central computer. The computer will have the necessary hardware and software to automatically switch functions as required, be it fax, voice, voice mail, or other message functions.

STATIONERY NEEDS

To establish your new business identity and facilitate the everyday aspects of marketing, accounting, and billing, you will need a variety of business forms. It is not necessary to get these forms right away; they are certainly not the essence of a sound practice. But if you are in consulting for any significant amount of time, forms can be a great time-saver and image enhancer. The effort and cost of preparing them are small in comparison with the potential benefits.

One way to obtain the forms you need is to find a good printer in your area, and ask him how he can implement some of the following ideas for you. If you go this route, first sketch out the design of the forms for the printer. The printer will then have a designer work your sketch into camera-ready copy and print the forms in quantity. This traditional method of obtaining forms gives you the highest quality, but the costs can be significant.

What items should you consider having printed? To list but a few:

- Letterheads
- Envelopes
- Labels
- Invoices
- Purchase orders
- Business cards
- Brochures
- Note pads
- Standard terms and conditions

An alternate way is to use your computer and laser printer along with form-making software. First design the form with the computer software. When it's time to send a form, fill it in electronically and print it out. Because you print only the forms you need, this is essentially a "forms on demand" system. It has the advantage that your forms are easily changed. If you change your address, for example, just modify that single line in the forms

Tip

In recent years, new specialty papers offering *coordinated preprinted color design* have become available. Just select matching letter sheets, envelopes, brochures, and cards from a variety of styles. They come in attractive color designs with a quality that would be difficult to achieve on your own laser or ink-jet printer. Then add your own text content and pictures, and print on a standard black-only printer. The results are considerably more appealing than what can be achieved with a monochrome printer.

software. (By contrast, when you get your forms from a commercial printer, modifying even one line means more design work, a new stack of printed forms, and greater cost. The pile of old forms is suddenly useless.)

The do-it-yourself approach to business forms also has disadvantages. You must buy the forms software and learn how to use it. Unless you spend a lot of time becoming proficient, your results will look amateurish compared with those of a commercial printer.

CREATING YOUR COMPANY LOGO

A logo is a small design that becomes a "visual icon" for your new company. In the eyes of customers, it becomes your *mark*—something they can visually associate with you. The best time to create a logo is before you order stationery and forms. Use your logo on all forms and promotional materials. Being consistent helps develop your company image.

Once designed, your logo becomes your trademark. A *trademark* is defined as "a word, phrase, symbol, or design, or combination thereof, that identifies and distinguishes the source of the goods or services of one party from those of others." (For more information on trademarks and how to register them, see *Basic Facts about Trademarks,* published by the U.S. Government Printing Office.)

Example 18 shows some effective logo designs. The idea is to base your logo on a simple geometric concept that encapsulates your company's name or initials. Ideally, your logo should project an image that is consistent with your business's service or product line.

EXAMPLE 18. Famous logos.

Suppose our consultant friend John Smith wants to create a logo. Where does he start? Well, John Smith Associates and its initials, JSA, are good first candidates. First, create a number of preliminary designs using an illustration program such as CorelDRAW!, Macromedia Freehand, or Micrografx Designer. These programs have extensive clip art libraries that make assembling designs very easy. For John Smith Associates, here are just a few possibilities:

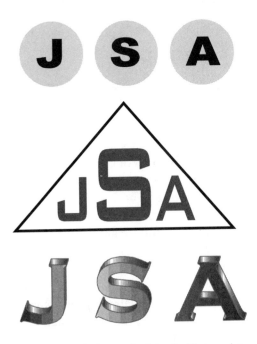

EXAMPLE 19. Logos for John Smith Associates.

MAKING YOUR COMPANY LOGO

- Make it simple. A logo is *your* "visual icon." Capture an essence, don't overdo it with a detailed illustration.
- Make sure it's legible.
- Design it to print well under a variety of conditions.
- Make it distinctive.
- Associate it with the theme of your company.
- Check that it's compatible with the image you are trying to project.
- Make good use of color and/or shading.

Creating an *original* logo is not an absolute necessity. If you find an attractive piece of clip art that captures the essence of your business, there's no rule that says you can't use that instead. Just make sure that the clip art is royalty-free. (And don't register it, because it's not *your* intellectual property!)

After considering the examples above, John Smith decided he wanted to use a small clip art image instead of a custom logo. The clip art more successfully conveyed the image he was trying to project. Example 20 shows how he integrated the design into his letterhead:

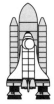

JOHN SMITH ASSOCIATES

2 Forbes Road
New York, NY 21225
212-435-6500   www.johnsmith.com

Consulting in Space Shuttle Thermal Design

EXAMPLE 20. Letterhead of John Smith Associates.

WHY YOU NEED BUSINESS CARDS

Clients need to have your phone number to call you, and the business card is the traditional way of accomplishing this. *Always* carry your business cards with you, even on trips to the tennis court or the supermarket. You can never tell where or when you will meet a prospective client or make a useful contact.

42 Bent Oak Trail
Fairport, NY 14450

716-223-4502

Harvey Kaye

Consultant in
Fluid Mechanics and Heat Transfer

HK Harvey Kaye & Company

EXAMPLE 21. My business card.

A business card (see Example 21) should have your name, address, and phone number, and a phrase under your name to indicate the nature of your consulting. Typical phrases might be "Consulting in Optical Design" or "Computer Hardware Engineering." Be careful if you use the word "Engineer." In some states, an individual must be a registered Professional Engineer (P.E.) before he can call himself an "Engineer" in promotional material. Check into this requirement before you have your printing done.

Your business cards will look better with your logo on them. Make sure that the logo is the correct size relative to the size of the card. It should neither dominate nor be dominated by the text information. Also, now you see why your logo should be *simple*. If it contains too much detail, it won't scale down properly to fit the standard 2-by-3.5-inch business card.

EXAMPLE 22. Business card with logo.

Keep the type size in proportion to the card. Use at most two different font types (such as Arial and Times Roman). In Example 22, the company name is in a large typeface for emphasis. Don't try to include too much detail in the card beyond the usual basics of company, name, title, company description phrase, address, phone, and E-mail address or Web site.

If you consult in more than one specialty, consider using different business cards for each one. I know many consultants who use multiple cards with good results.

HOW TO WRITE A DYNAMITE BROCHURE

Your brochure is probably your most important sales tool. It should be composed with care and intelligence and printed on a top-quality heavy-weight paper. I personally prefer paper in a light pastel color with contrasting ink.

For the beginning consultant, a one-page brochure is sufficient. It should contain the following:

- Your company name, address, and phone number.
- A statement of the services your company offers. List the specialized technical areas in which you are competent.
- A capsule description of your background. Include items that enhance your image as a consultant. Leave out extraneous details.
- A statement of the *application areas* in which you have experience. In this context, "application areas" refers to particular industries and market segments rather than technical subdivisions. For example:
  - Software for the home appliance industry
  - Military qualification procedures
  - Seismic testing for the nuclear power industry
  - Circuit design for the audio electronics market
- The benefits that the client will gain by using your service. (Recall from Chapter 6 that clients don't "buy" on the basis of *your* features, but on benefits that *they* will receive.)

The layout of your brochure should be visually attractive. Consider adding graphics or photographs to enhance its appeal. Before printing, have knowledgeable colleagues check your final draft for spelling, marketing slant, and general appeal. A sample brochure is given in Example 23.

I always enclose my brochure with proposals that I submit to clients. It adds a measure of credibility to my proposal and avoids the effort of having to describe the general aspects of my consulting practice to new prospects.

Specialized Consulting in HEAT TRANSFER FLUID MECHANICS

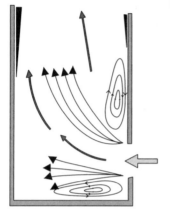

Areas of Expertise

We have over twenty-five years' experience in the following technical areas:

- Conduction, convection, radiation
- Compressible and viscous flows
- Supersonic and hypersonic heating
- Turbulent flows
- Chemically reacting flows
- Hydraulic networks
- Two-phase flows
- Boundary-layer applications
- Ablation and transpiration cooling
- Thermodynamics and cryogenics
- Microelectronic junction thermal analysis
- Thermal packaging and cooling of electronic assemblies
- Finite-element programs
- Thermal stress analysis
- Reliability assessment
- Software development in FORTRAN, PASCAL, and BASIC

Capabilities

Harvey Kaye & Company offers a practical systems approach to your problems in heat transfer and fluid mechanics. Let us help you with:

- System configuration
- Trade-off studies
- Preliminary design
- Testing
- Analytical and computer methods
- Independent assessment of problems
- Research and literature surveys
- Computer software development

Industrial Applications

Commercial appliances and products
Missile radomes, fin control surfaces, and internal components
Heating, ventilating, air-conditioning
Chemical process equipment
Infrared optical systems
Microelectronic devices
Laser beam–path conditioning systems
Nuclear pressure vessels and piping
LNG technology
ATE systems
Computer systems
Gas turbine technology

Why Choose Harvey Kaye & Company?

Wide experience in heat transfer and fluid mechanics. In fact, we may have already solved "your problem."
Practical understanding of the functional, structural, manufacturing, cost, quality, safety, and management interfaces.
Excellence in written reports and oral presentations.
We listen to determine your needs.
Reputation: Since our start in 1976, we have developed a long list of satisfied customers.

HK Harvey Kaye & Company

42 Bent Oak Trail
Fairport, NY 14450
716-223-4502

EXAMPLE 23. Brochure.

WHAT IS A CONSULTING RÉSUMÉ?

Clients often request that you submit a copy of your résumé along with your proposal or quotation letter. They simply want to better understand your technical experience and qualifications. However, if you give them a standard résumé, it may look like you are trying to solicit an employment offer. Standard résumés usually state (in effect) that you are looking for a new position. How do consultants get around this obstacle? They send in a *consulting résumé*, which makes it transparently clear that they are already employed and *not* looking for a new position.

A consulting résumé is similar to the standard personal résumé, except that the statement of "job objective" at the beginning is replaced by a short "qualifications" statement. (See Example 24.) Also, the first sheet should be on your company letterhead. The consulting résumé should be longer than the personal résumé. Ideally, it should stretch to three pages. It should give sufficient description of your work so that clients can determine the depth of your experience and your suitability to their applications. Use specific project names and industry buzzwords. Highlight your accomplishments. The résumé should also include professional affiliations, educational background, recent short courses, and a list of your professional publications and patents.

Bias your consulting résumé to make yourself look like an expert. Present yourself in the most optimistic light. Emphasize all relevant experience and credentials. Highlight everything that supports your competence and expertise. Consulting résumés need not follow the "full disclosure" standard that applies to direct-employment résumés. Therefore, omit or downplay any items that portray you unfavorably or that distract from the image you wish to project.

With your laser or ink-jet printer, you can create *exactly* the résumé you want, tailored to each customer's particular expectations. These *résumés on*

<u>Warning</u>

Do not include details of your personal life, such as hobbies, marital status, and religious preferences, on your consulting résumé. *Never* put your rate on your résumé or, for that matter, on any of your printed forms. You will probably have to quote your rate in a separate proposal or quotation letter anyway, and each of your clients will have its own required format for submission of this information.

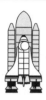

JOHN SMITH ASSOCIATES

2 Forbes Road
New York, NY 21225
212-435-6500   www.johnsmith.com

Consulting in Space Shuttle Thermal Design

CONSULTING RÉSUMÉ OF JOHN SMITH

QUALIFICATIONS

Twenty years' experience in vibration analysis and testing, structural dynamics, and shock-resistant design. Strong skills in mechanical engineering design, CAD, computer programming, vendor specifications/negotiation, and oral/written presentation.

EXPERIENCE

Consultant, John Smith Associates, 9/82–Present: Management of consulting firm with overall responsibility for marketing, contract negotiation, and payroll administration. Consulting in the areas of vibration analysis, testing, and design. Clients and accomplishments include:

General Electric Corporation, Corporate Research Center: Development of a unique computer program to assess vibration and shock loads on a lunar landing vehicle.

Digital Equipment Corporation: Study to determine the feasibility of using artificial intelligence techniques to automate the design of vibration dampeners for microelectronic testers. This study saved Digital $250,000 per year and won its "Best Consultant Report of the Year" award.

United States Steel: Design and analysis of a novel vibration snubber for robotic assembly machines. Responsible for design specification, engineering trade-off studies, and detailed mechanical design. John Smith Associates is coholder of the patent granted for the design.

Honeywell: Design and analysis of vibration mounts for the SYM transport proposal. Definition of system requirements, cost/performance evaluation of candidate systems, and equipment selection for the chosen system. John Smith Associates' presentation of the vibration design to Honeywell's customer was instrumental in winning the contract.

Direct Positions

Fair Haven Designs, Inc., Manager of Research, 4/78–9/82: Responsible for all aspects of the company's funded R & D, including client contact, proposals, coordinating projects, and presentation of reports. Areas included snubbers, computer-aided vibration monitoring, and shock isolation design services.

EXAMPLE 24. Consulting résumé.

Tamara Associates, Senior Engineer, 8/75–4/78: Responsible for obtaining and managing projects in vibration and dynamic stress analysis of nuclear power plant components. Elastic-plastic analysis of Low Down unit I steam nozzle using finite-element codes. Developed an original statistical determination of seismicity levels for the Wee Willie nuclear station in Burma.

Air-O-Vac Corporation, Engineer, 6/73–5/75: Computer and analytic studies of vibration modes of an IBM microchip assembly undergoing transient thermal cycling. Produced an original FORTRAN code to determine the displacements and material response.

EDUCATION

University of Rhode Island, B.S. in Mechanical Engineering, 1973
University of Massachusetts, M.S. in Mechanical Engineering, 1975

PROFESSIONAL AFFILIATIONS

American Society of Mechanical Engineers
ASME Committee on Computer-Aided Vibration Analysis
New York Registered Professional Engineer #1244

BIBLIOGRAPHY

"A New Method to Derive Seismicity Levels," *asme Journal of Vibration,* November 1978.
"Novel Way of Predicting Vibration Modes in 7765 Microchips," *IBM J. Research and Development,* October 1975.
"Survey of Computer-Aided Vibration Design," *Design News Today,* June 1991.

PATENTS

"A novel vibration snubber for robotic assemblies" US Patent 35,789,456
"Shock isolator mounting for reduced-G operation" US Patent 33,543,987

TEACHING

ASME Short Course on Vibration Analysis, at Annual Meetings, 1990–1997.

---

demand are especially useful if you market to different specialties or industries. For example, suppose you have extensive experience in the electric power and aerospace defense industries, which are not particularly synergistic. Clients in one industry may frown on your experience in the other. Simply create *two* résumés, each one custom-tailored to its intended audience. Highlight experience relevant to the featured industry and downplay references to the conflicting one.

WHAT IS A COMPANY BUSINESS DESCRIPTION?

Some clients want to know your business a little better before they will give you a contract based on your proposal alone. Other clients (including many government agencies) may refuse to put you on their *approved vendor list* until you supply certain information. The standard format for providing this information is the *company business description*. This document should be furnished in a thin binder that contains your brochure and consulting résumé. In addition, it should include

- A description of your company. What is the form of business ownership: corporation, sole proprietorship, or partnership? What is the management structure? When was your company founded? How many employees do you have?
- A description of your office location and facilities. Mention any special equipment that enhances your image in the client's eyes.
- The disciplines and technical areas that you cover.
- A list of previous clients and projects. (It need not include every client and every project. Leave out the ones that bombed.)
- Consulting résumés for key employees and partners.
- Company billing rates and overhead structure.
- Benefits offered to your employees and yourself.
- Certification that you comply with all federal and state regulations and tax laws.
- Your billing procedures.
- The business insurance coverages that you maintain and the name(s) of the agency (or agencies) handling them.
- Policies for hiring, firing, drug testing, verification of credentials, and assignment of patent rights.
- Security clearance information.
- Bank references.

The typical company business description is ten to fifteen pages in length. Since relatively few clients ask for them, I would not make one up until the occasion demanded it. This report contains private information you definitely don't want circulated. Stamp the bottom of each page CONFIDENTIAL. Place a notice on the front page:

> Do *not* circulate! This information is private. It is to be used
> *only* for purposes of business reference and evaluation.

CHAPTER 15

Goals and Planning for the Future

PLANNING: ENGINEER YOUR OWN FUTURE

Goal formulation, planning, and self-evaluation are essential management functions of a well-run business. If you are a direct employee in a large firm or university, you have probably not been exposed to these functions. Technical experts are usually busy doing wholly technical work, leaving others in the firm to do the management tasks.

Beginning consultants often fail to understand that they now must assume the responsibility for managing and directing their own business and that these tasks are vital to their long-term success. The consequences of *not* formulating goals, planning, and evaluating are the same for both large corporations and individual consultants: stagnation, complacency, and diminished opportunities.

If you have already implemented a marketing plan similar to that described in Chapter 6, you doubtless realize that planning means *choosing* certain objectives over others. Planning focuses your efforts. Without targeting the kind of business you want, you will end up with a set of unconnected projects that lead nowhere. As a result, you will find yourself in an unfavorable strategic position. Management consultant W. M. Greenfield explains, "A practice is not just a bunch of projects." Greenfield reflects poignantly on learning this lesson:

> [Without] planning or goal setting,... I could exercise no control over what projects came my way.... Because I could not target my efforts, they remained scattered.... I could not really build a business, although I could and did add more work.... I felt out of control all the time and at the mercy of pure chance.

This is a truly rotten feeling. (*Successful Management Consulting,* Prentice Hall, 1987, p. 138)

Planning also helps you eliminate wasted motions, motions that dilute your primary business thrust and sap it of energy. With strategic planning, you can place your efforts in those areas where the payoffs are the highest.

Without targeting the business you want, you'll end up with a set of unconnected projects that lead nowhere.

Chapter 6 discussed marketing plans in particular. Here, we shall consider more general plans dealing with career choices and directions. This latter kind of planning need not be elaborate. In the following pages, I'll outline the simple steps.

The most important preliminary in career planning is developing an understanding of your present situation. The easy part of this involves learning your strengths, your weaknesses, and which resources you can draw on. The more difficult part is coming to understand your personal history in terms of the previous choices you made.

Everyone comes to the planning process with his or her own hidden biases and agendas. The psychological function of these hidden biases is to misrepresent your situation to yourself. This way, you can do what you naively hope is "right" for yourself without having to face certain less-than-wonderful facts. To understand your present position with any degree of insight, you will have to be very honest with yourself.

- Examine your attitude toward your technical field. How do you *feel* about it? Are you mildly interested, neutral, or very excited about it? Is this where you want to commit your energies? If not, what really turns you on?
- Look at your employment history. How did you reach this juncture in your life? Is your involvement with your technical specialty totally accidental? Have you been laid off or fired in the past? Are you happy with direct employment, but can't find a "good" job no matter how hard you try? Do you fit into the corporate mold? Are you advancing as you had hoped? Are you disillusioned — or thrilled about your prospects for the future? Do you even *want* to advance in your field? Are you unhappy about your employment track record? What decisions did you make that turned out right? Wrong?

- Scrutinize your personal life. How do you handle risk? Are you at your best when you're autonomous or when you're being closely supervised? Where have you placed your extra time and energy? Have you spent some of your own resources on professional advancement, or is it all directed into maintaining the summer cottage or taking a six-week ski vacation every year? Are your loved ones supportive of your new endeavors? Are they in a position to be supportive of your venture into consulting?

Until you have the courage to answers these questions honestly, you will have great difficulty setting worthwhile goals and making effective plans to reach them. Instead, you will wind up with half-baked plans that don't work because you neglected the basic psychological facts of your situation.

WHAT ARE YOUR GOALS?

Let's talk about goals first. Our minds are easily deceived when it comes to the matter of goals. Often, our goals are not what we think they are, but merely imitations of other people's goals. Your own goals may remain unknown to you, even as you read this!

It is important to realize that we choose from a multitude of goals. Life is more than professional work, and each one of us holds many goals simultaneously. Each of these simultaneous goals has a different priority according to our personal value system. Further, goals exist within *environments* and are subject to *constraints*. The environments and constraints are significant parts of the goal formulation and have a large influence on how easily the goals can be met. It is helpful to develop awareness not only of your goals and constraints but also of the priority you assign to each.

A man walked into a bar, trailing an extremely foul odor behind him. He sat in a corner all alone. All evening he sat there, nursing his beers. Finally, a lady felt so sorry for him that she held her nose and went over to him.

"Yes," he said, "it's always like this; I'm always alone. Nobody talks to me, nobody even comes near me, because of the smell. It's my job at the circus. I pick up the droppings behind the elephants, and the stench is terrible. I have no friends, and my life is terrible lonely."

"In that case," she asked, "why don't you quit your job?"

"What!" gasped the man. "And give up showbiz?"

In the present context, the *goal* you may be evaluating is becoming an engineering consultant. The *constraints* are your financial assets, your financial liabilities, your professional reputation, your energy level, and the impact of competing personal goals. The *environment* may include such factors as a healthy national economy, a stable family situation, and high demand in your technical field. And your priority ranking may be: family first, work second, wealth accumulation third — and showbiz last!

ASKING THE RIGHT QUESTIONS

A married man went to see a specialist in internal medicine. "Doctor," he exclaimed, "I'm really desperate! Last week I came home from work and found my wife in bed with another man. When she saw me, she started weeping and begged me to forgive her. So I said, 'Well, let's have some coffee...'

Then, the other night, the same thing! She was in bed with my neighbor! I took a pistol to kill them, but she cried and cried, so I said, 'Well, let's have some coffee...'

Last night, I couldn't believe it! The same thing again! She promised me it was the last time. She said she'd never do it again, so I said, 'Well, let's have some coffee...'

"Doctor, I am very upset! Is it OK to drink this much coffee?"

In goal setting, asking the *right* questions is essential. Irrelevant questions serve only as distractions. Yet most beginning consultants are preoccupied with issues that are peripheral to their success as consultants. For example, I frequently hear prospective consultants saying:

- "I can't wait to get out of this bad situation." (The focus is on consulting as a means of escaping an annoying short-term situation rather than as a long-term strategy.)
- I bet my present employer will give me a large contract when I leave, because the project I am working on will fall apart when I go." (The focus is on the employer's problems rather than the consultant's.)
- I can't wait to get my Cadillac so I can deduct it on my taxes." (The focus is on the potential financial reward rather than the activity itself.)

Sharp questions about your goals force you to sift out the issues, define your constraints, and prioritize in a way that will indicate clear action. In the context of setting career goals, ask yourself:

- How do I want to earn my living?
- What do I have to do to attain this goal?
- What must I give up to attain this goal?
- Do I have the will to do it?

THE HIERARCHY OF BUSINESS GOALS

I have observed that business goals follow a hierarchy based on two dimensions: survival versus comfort and today versus tomorrow. The order of the hierarchy is

1. Survival today
2. Comfort today
3. Survival tomorrow
4. Comfort tomorrow

In starting a business, the emphasis is on surviving for the short term. Under these circumstances, long-range planning is a luxury. Your first contracts will probably involve solving other people's problems 100 percent of the time, and you may find these problems of little personal interest. Your immediate goal is to build up a cash reserve and a reputation for being of good service. You are gaining business experience and are learning to stay afloat in a sea of tax forms, proposals, contracts, and accounting books.

As you become more established in consulting, your goals will probably evolve along the steps of the described hierarchy. After a while, you will find yourself spending more time exploring professional activities that tap your creative powers and give you the feeling that you are growing. As your client base grows, you will be able to become more selective. You will be able to choose the customers that are best for you in terms of *where their projects lead you technically.* You will be able to lean into your interests. You will also be able to invest newly generated resources in equipment and efforts that give you a competitive edge or allow you to do your work more efficiently and comfortably.

STRIVING TO SURVIVE

The beginning consultant is often faced with strategic dilemmas. These strategic decisions are trade-offs on the dimension of survival today versus laying a better foundation for tomorrow. For example, should you divert a significant portion of today's revenue stream to keeping abreast of newly expanding areas that may provide tomorrow's business? Or should you ignore the new technologies until they are absolutely necessary? That way, you save time and money, but you may find yourself in a poor strategic position two years later.

Consider the case of my friend Alan, who in 1995 was trying to change from his role of DOS FoxPro database programmer. Alan predicted that DOS FoxPro would soon become obsolete. Most clients were looking to the newer Windows database programming platforms, such as Visual Basic. After much deliberation, Alan made a strategic decision: he would take two months off and study Visual Basic. The goal he set for himself was to get a *credential* by passing the Microsoft Certified Professional Exam in Visual Basic. He put his plan into action, which meant that he was receiving no income while he was studying.

Two months later, he passed the exam and went about finding a contract to do Visual Basic programming. As luck would have it, the second week into his marketing efforts, a client asked him to take on a project in DOS FoxPro programming! Alan was faced with a difficult judgment call: Should he take the FoxPro work right away or wait a little longer and hold out for a contract in Visual Basic?

Alan was able to pose the decision as a trade-off in which both alternatives offered pluses and minuses.

Alan's decision depended on the fact that he had planned for a slow start in this new area. He had saved enough cash to comfortably wait while he doubled his marketing efforts in Visual Basic. Sure enough, two weeks later, he landed a contract to design a database system with Visual Basic. I am happy to say that his business has flourished as a result of his strategic plan and his courage to see it through.

Tip

As a general policy, if you lack the financial resources to wait, accept the best work you can find, realizing the compromise implicit in the trade-off situation. Then work very hard to improve your position *strategically* once your revenue stream is flowing again. You might do this, for example, by continuing to market for the new area, by creating more credentials, or by making more professional contacts in that area.

Take FoxPro Contract Now	Wait for a Visual Basic Project
Revenue stream (improves financial position)	No immediate revenue stream (hurts financial position)
Revenue stream permits allocation of some resources to marketing Visual Basic after current project is complete	Don't know when project will come (uncertainty factor)
Old technology (dead-end; hurts marketing strategic position)	New technology (better strategic position for tomorrow)
Lost opportunity penalty (taking FoxPro now precludes other choices for x months)	Time to do more marketing in Visual Basic and create more credentials

PLANNING FOR TOMORROW'S SUCCESS

When your business is well established, consider expanding your base of operations to make survival tomorrow easier. The forms of expansion are innumerable, but you might start with the following:

- Enlarge to multiperson operation: add employees or associates.
- Specialize more narrowly (vertical approach): build reputation in a narrow field by publishing articles, writing books, participating in conferences, etc.
- Diversify technical fields (horizontal approach).
- Slide into a new technical area gradually.
- Manufacture a product.
- Buy equipment that allows you to enter new market areas.
- Invent products.
- Add sales reps and branch offices.

I have used strategic planning myself to gradually change the focus of my consulting practice. For the past ten years, I have been intrigued by personal computers. In the early eighties, they were little more than curiosities and toys. As their capabilities evolved, I found myself spending more and more time with them. With recent advances in microprocessor technology, personal com-

puters have become capable of serious technical and professional development.

At first, I did not have a goal, so I just "played" with the software. As I got more dedicated, I joined the Rochester PC Users Group, for which I served as vice president for four years. I made dozens of useful business contacts in this capacity. Around this time, PC programming languages started to emerge as the development platforms of choice for technical work. It was great fun to program in these new languages—they were so much more powerful and user-friendly than the mainframe languages I grew up on. At this point, I wanted to know how I could gradually slide into this new field.

In looking at my career history, I asked myself where I was making a contribution. The answer was "Using programming for engineering purposes." Even from my grad school days, my strength has always been in programming engineering applications. Since I already had credentials in this narrow subset of programming, it seemed logical to repackage myself as a programmer in general. I started writing articles for PC journals about technical database applications. I spent significant amounts of my spare time creating PC programming applications to serve as a portfolio.

After five years, my credentials were strong enough to begin marketing myself in the programming field. I have met with considerable success so far. I still consult in heat transfer and fluid mechanics (my old fields), but now I also offer a new "product" that is more closely aligned with the way my interests have evolved over the years.

Pursuit of the "tomorrow" goals may even include business activities other than technical consulting. That is, consider contingency plans in the event that your area of consulting dries up or that your interests gradually shift to other professional areas. These plans can be explored at a low level of effort while you are busy with your present work. Part of this planning process is preparing for events that can challenge your present prosperity. Believe me, dry periods and shifts of interest *do* occur!

Any engineer who was working in 1970, for example, was acutely aware that a sudden change in national priorities resulted in a near-depression situation for tens of thousands of engineers. I recall a *Time* magazine article describing how the fortunate among the unemployed engineers started pizza parlors, while the rest washed dishes at local restaurants and pumped gas. Strong cyclic demand is not just a recent phenomenon, though. Edwin T. Layton's book *The Revolt of the Engineers* (see the Suggested Reading List) documents the boom-or-bust nature of engineering in the United States over the last 150 years.

EXPANDING YOUR PRACTICE WITH ADDITIONAL WORKERS

Successful consultants are often offered more work than they can handle alone. When this happens, they have a fundamental choice: they can turn the work down or they can expand their practice by adding staff members. The first few times you are offered extra work, you may not be ready for it. You may never have considered the possibility of expanding your operation. However, if you are *consistently* offered more work than you can handle, expanding your business with subcontractors is a wise and profitable solution. As discussed in Chapter 13, using subcontractors instead of employees reduces the tax, accounting, and administrative burden. Furthermore, it allows you to shrink your operation back down if, for some reason, you decide that multiperson operations are not for you.

Developing a multiperson operation can be very lucrative. Suppose the contract you are negotiating calls for the efforts of three technical consultants. All you have to do is find two fellow consultants who will subcontract to you to perform the additional work. Suppose you charge the client $75 per hour for their services. You offer to pay your subcontractors 80 percent of this amount,[1] so that you gross $15 per hour per subcontractor. That's $1,200 gross profit per week!

<u>Warning</u>

There are two cautions to observe in running a multiperson operation. First, pay greater attention to cash flow considerations. You are now responsible for paying others as well as yourself. Your monthly payroll may become so large that dipping into personal financial reserves may no longer be a reasonable solution to cash flow crunches. Second, make sure you exercise *subcontractor tax releases* with all subcontractors. As mentioned in Chapter 13, check that your subcontractors qualify as bona fide independent subcontractors and are not actually common-law employees.

---

[1] The prime contractor's cut usually ranges from 10 to 30 percent. A friend who is an experienced businessman once told me, "Never take less than 10 percent on a business deal, even if your only responsibility is to send a simple invoice and write a single check." It took me many years to understand the wisdom of this advice.

Your *net* profit is determined by subtracting the expenses of carrying the subcontractors. Such expenses include increased insurance rates, workers' compensation, and the costs of administering payroll and personnel. Although these expenses are far from negligible, net profits typically run 75 percent of the gross.

Of course, managing the efforts of your subcontractors also requires some effort on your part. "Full cost recovery" accounting would charge the hours you spend in management activities against your net profit. Even with this more conservative view, the income potential is still very high.

The beauty of using subcontractors is that your contractual liabilities are very small. If the client suddenly terminates the contract or dismisses one of your workers, you don't have to find work for him or pay unemployment taxes.

Expanding with subcontractors is a low-risk way to enhance the credibility of your company. It opens doors to new opportunities. However, when your business reaches an even higher level, you may be ready to assume the risk of becoming an employer. Although this entails more taxes, bookkeeping, and management, the rewards are also greater. Employees offer loyalty and continuity that are not possible with a subcontracting operation. For many consultants, running a company staffed with their own employees is a dream come true.

MEET THE PROS UP CLOSE AND PERSONAL

Janice Rivenburg is the president of Rivenburg Associates, Inc., a computer consulting services company. Her academic background is in business administration with a concentration in data processing. Janice's first few jobs involved headhunting for computer programmers. By 1986 she  wanted to start her own computer consulting company. Since she lacked experience, she went to work for a small contract house to learn the ropes. She was so successful at matching computer consultants with contract positions that her commissions exceeded the owner's expectations of what a recruiter should make! The owner "rewarded" her by cutting back on the commission rates he had promised her. Janice saved her money and made plans to start her own business.

She founded Rivenburg Associates, Inc. in 1991 as an S Corporation. Janice started with no employees, but by the end of her first year, she had six. She now has many more people and is growing at a steady rate.

Although Janice has skills in computer programming, she plays into her strength, which is marketing and dealing with people. She does all her marketing by networking and word of mouth. Computer consulting is an extremely competitive field, and Janice maintains her competitive advantage by keeping her overhead low and selecting her employees judiciously. This way, she is able to offer her workers a higher percentage of the gross billing rate. Further, she allows employees to "buy" only those benefits that suit their needs, in contrast to most competitors, who force workers to take (and pay for) the whole benefits package.

Janice's other competitive edge is her attitude toward clients and employees. "People want to be treated with honesty and respect. I have an open-books policy about rates. Every employee of mine knows what the client is being charged for their work." Every client is presented only with candidates that Janice carefully screens for the job. Janice finds this matchmaking role both challenging and rewarding: "The client's time is very important. Other consulting companies waste it by sending out poorly matched or unqualified candidates. When I tell a client that I have a match, they listen more attentively."

Janice advises consultants who are considering multiperson operations to treat both employees and clients fairly. Get to know them well; spend enough time to understand what they want from the deal. Building trust is an essential ingredient in this formula, because your reputation follows you. Moreover, Janice says, "think strategically! To save wasted effort, aim only at the best targets at the most opportune moment."

COMPLEMENTARY ACTIVITIES

Teaching at the university level is a wonderful way to enhance your visibility and credibility. Many schools now augment their regular faculty with visiting and part-time appointments. Hiring outsiders benefits schools by maintaining low instructional costs and by infusing real-life experience into the curriculum.

The credentials you need to teach depend on the position. A master's degree and several years' relevant work experience are the usual minimum. Beyond this, teaching requires a scheduling commitment from the consultant. Failing to show up at class because you have been called out of town on a hot consulting project is unacceptable. You must reserve the time not only for classes, but also for preparing, administering, and grading exams; correcting homework; and other academic chores.

Teaching is great for consultants. It sharpens your ability to make presentations and promotes confidence in public speaking. Classroom interaction helps you to think on your feet in handling technical questions. Teaching consolidates and refreshes many technical subjects you may have forgotten over the years. Further, you can make useful consulting contacts through students and other faculty.

Working with students is fun and rewarding in its own right. You can explore your own technical approaches and share them with your classes. You can be as creative as you want; most schools encourage you to make your own classroom demos, learning aids, and special term projects.

Teaching also provides additional income. However, because most outside teaching appointments are part-time, the salaries and benefits are usually low. For this reason, I recommend teaching more as a valuable experience than as a moneymaker.

Giving seminars is another activity that complements consulting. Once you have selected your topic, reserve a meeting place and advertise for attendees. Herman Holtz's book *Expanding Your Consulting Practice with Seminars* (see the Suggested Reading List) shows you the details of seminar format and pricing. Holtz encourages consultants to structure their seminars to be synergistic with their consulting business. You may break even on the seminars but gain income overall as a result of the new consulting prospects you generate.

Writing books has been a traditional sideline for consultants and academics. Even in today's information-overloaded environment, technical books and monographs are powerful credential builders. Before embarking on this literary endeavor, you should consider three challenges:

- Writing the book itself, which may take a few years. Unlike articles, books require an enormous amount of time. Spending a thousand hours to write a two-hundred-page book is not unusual.
- Finding a publisher, which is easier said than done. (Try to line up the publisher *before* you write the book!)
- Getting a decent financial return on your invested time. Because of their limited market, specialized technical books rarely attain best-seller status. Most authors report cumulative royalties to be less than five thousand dollars.

MEET THE PROS UP CLOSE AND PERSONAL

Andrew Brust is the president of Pro-
gressive Systems Consulting, Inc., a New
York City consulting firm that develops
Internet, client-server, and other custom
business applications in Visual Basic for
banks and small businesses. Andrew has
a B.S. in computer science and spent the

first seven years of his career working as an in-house computer support
expert and a database developer. In 1995 he started his consulting firm
and soon landed a large contract from a software training company. In
two years his business had grown to three employees.

About the same time that Andrew started his business, he realized
the need to get broader exposure. He decided to write technical articles
for the *Visual Basic Programmer's Journal.* After writing a number of
articles, he was asked to be a contributing editor. Andrew also gives
seminars at the annual VBITS Visual Basic conferences. These seminars
are a natural for Andrew, because they draw on his daily experiences
as a programming consultant.

As a result of this exposure, Andrew's business has increased substantially. He says, "Writing gives you a reputation that precedes you.
In some cases, the contact has never even seen me at a seminar, but just
sees the advertisement for the seminar and calls for consulting help. In
other cases, the contact sees one of my programming articles and my
bio and calls to inquire about consulting services."

Andrew's advice to other consultants is "If you write well, don't be
afraid to approach magazine editors and ask for the chance to submit an
article." He also emphasizes that consultants need to understand that
writing books and articles is not intended to generate revenue, but to create credentials. "Think of it as getting paid for your own advertising."

Further, in giving seminars, Andrew advises to prepare very thoroughly at the beginning. "Make sure you know your lines really well.
It's easy to underestimate the time and effort required to create a good
presentation." For Andrew, developing a one-hour presentation takes
as much as eighty hours of research, sample-code preparation, and
practice time. There's no room to be dry or awkward, so Andrew
always tries to punctuate his material with real-life examples, anecdotes, and a touch of humor.

For many consultants, the gains in enhancing reputation far outweigh the monetary concerns. By targeting their subject and audience carefully, authors can become "experts" in a relatively short time.

Publishing your own newsletter requires less effort than writing books and involves less risk. Yet, they are an excellent way to enhance your credibility as an expert in your technical specialty. Newsletters inform readers about the latest technical developments in a specialized field. They focus on interpretation of trends and hot news items about which the reader is vitally concerned. Many newsletters are not afraid to "go out on a limb." They offer the writer's own vision and advice on specific issues to help readers make better decisions.

The nice thing about newsletters is that you can desktop publish them yourself with relatively modest computer equipment. You have great control over the content. And you can use your creativity in soliciting guest columns to provide variety and spice.

Newsletters offering rehashes of widely available information don't survive in this competitive field. On the other hand, if you have a steady pipeline to hot tips, your newsletter may grow into a lucrative side business. Annual subscriptions for a typical four-page monthly range from $40 to $1,000. Suppose you charge $100 per year for your newsletter. You would need only 1,000 subscribers to attain a $100,000 business level.

For more details on newsletters, I recommend *Starting and Running a Successful Newsletter or Magazine* by Cheryl Woodard, published by Nolo Press. Before I leave this subject, here are some additional guidelines that may help:

1. Stick with your expertise. Publish only in those areas in which you are personally knowledgeable.
2. Target the audience that's right for you. The content, length and depth of articles, and slant must consider the very people you want to reach.
3. Be a newsbreaker. At least some of the content should cover inside or late-breaking news. Don't carry "timeless" content only.
4. Cut your costs by taking on advertisers and by using the latest technology for printing and distribution.
5. Keep quality high. A few articles can be "filler" material, but the majority must really say something.
6. Leverage your newsletter's credibility by asking notable industry sources for interviews or to contribute articles.
7. Understand the competition. Investigate what they are doing and consider how you can apply their best ideas to your own efforts.

The founder of the modern newsletter was Willard M. Kiplinger, who started the *Kiplinger Washington Letter* in 1923. Kiplinger had just one new angle in creating his newsletter, but it was a great one: Give readers the inside scoop in four pages of short, easy-to-digest articles. He targeted his newsletter at business leaders who were eager to learn the inside story but who didn't have the time to study each topic in depth. This formula has worked well, for the *Kiplinger Washington Letter* has become one of the country's largest newsletters, with 330,000 subscribers paying $63 per year.

Another complementary activity is becoming an agent for a vendor or manufacturer. The prime requirement for this is a strong interest in sales. The contacts you make as an agent will lead you to new consulting clients. For example, suppose you are an expert in air-conditioning. You might align yourself with an air conditioner manufacturer as a local rep. You not only generate income from the sales activity, you also generate new leads for consulting.

In deciding whether to become an agent, ask yourself: How closely does the activity align with my consulting? Will it enhance my circle of contacts? Will it *limit* my opportunities by virtue of any noncompetition agreements I must sign? How much will my activity as an agent affect my ability to schedule consulting hours?

LET YOUR CLIENTS PAY YOUR DEVELOPMENT COSTS

In many situations, your clients will pay you to develop new technology for them. This is not only exciting and fun; the new technical capabilities you develop will lead to new business opportunities. The secret is to find projects that combine your present talents with the new technical areas you wish to develop. Your clients may be hard-pressed to find people who are expert in both areas, so your chances of landing such work are better than you think, providing you do a little homework.

Some years ago, Anne, a friend who works at Digital Equipment Corporation (DEC), was telling me about her exciting work in artificial intelligence (AI). She was developing an *expert system* program that configured the computers DEC manufactured. In the course of casual conversations, she explained how artificially intelligent systems could completely change the way engineering design was done.

I borrowed a few of her books on the subject and started to read. It wasn't long before I was searching out additional books and articles on this subject, as I thought there might be an application that tied this new area to my own specialty.

My reading on AI prompted me to ask: Could AI be used to design the cooling system of electronic cabinets? At the time, electronic manufacturers were doing the same kind of design over and over for their cooling systems. Could the process be automated and improved by the use of expert systems?

One important characteristic of the thermal design process was that a number of design parameters had to be considered before a clear design direction could be determined. The way in which they were considered by the designer was the same way an expert system used its *rule base* of expert knowledge. I conjectured that an AI program might be able to look at the parameters for each cabinet and make its own judgments about the direction in which the design should proceed. In the case of a system with low thermal power, for example, the "rule interpreter" would use the rule base and determine that a natural convection cooling system should be used. The same cabinet with a high-power system would be cooled by a forced air or liquid cooling system, depending on other input information and intermediate decisions made by the program.

I knew an electronics company that would be interested in such an expert system if it were available. It produced a large number of systems, all of which needed to be cooled, but each cabinet had custom options that changed the cooling parameters significantly. A perfect opportunity!

I posed my ideas about this application of AI to the person in charge of that company's mechanical system development. I outlined the potential for saving dollars and manpower, and showed how the expert system fit in. Since I was learning about expert systems myself, I proposed a feasibility study, in which I would become more acquainted with the AI systems on the market and the state of the art. I did not promise to deliver an expert system. I was looking at the new technology to see if it could be applied to that company's need.

I wrote a three-page letter describing my idea, and appended a recent journal article on the fundamentals and applications of AI. This article emphasized that every company, if it wanted to stay alive, had better learn to apply this new technology!

I subsequently did a two-month feasibility study on expert system methodology for the application. The client was only too glad to have me work on this problem and offered to pay for expenses such as books as well!

WHY ONE-YEAR PLANS ARE BETTER THAN FIVE-YEAR PLANS

The key to keeping on top of your own development is to plan and evaluate regularly. Don't let your tactics (your reactions to the inevitable business crises occurring on a day-to-day basis) take the place of strategic planning. Devote two days a year to short- and long-term planning. Although it is important to have a general idea of where you are going, as might be reflected in a five-year plan, it is much more useful to have a well-conceived and detailed one-year plan. Five-year plans do not have a sense of urgency, and the common reaction to them is to ignore them completely.

As part of your one-year plan, set yourself a number of specific action items that will bring you closer to your long-term goals. Try to envision what efforts these specific tasks will entail on your part, and ask if they can reasonably be done. It is crucial that the specific tasks are achievable in the time frame you have established. Setting goals that cannot be achieved within the time frame is self-defeating.

 Tip

For my own planning sessions, I go to the mountains or the beach, away from any crowds, and bring a set of motivational tapes. (There are hundreds of these on the market, dealing with inspiration, planning, and reaching your goals.) I play the tapes and listen to them intently; this is my conference.

I start by asking myself the big questions:

- Where have I been this past year?

- Where am I now?

- Where do I want to be a year from now?

Then I slide into practical strategies for the next twelve months. I note technical areas that look really exciting and find ways to become more informed about them. When I return, I purchase the books, courses, and materials that will bring me up to speed in the new areas. I also extend my hold on my present specialty by figuring out the technical papers I should write, the journals I should read, and the contacts I should make.

WHAT TO DO WITH GOALS THAT NEVER GET ACCOMPLISHED

Sometimes, things don't work out exactly the way you planned them. That's life. But what happens if your planning sessions show that you *consistently* set goals that never get accomplished? In this case, your "life" may be telling you something:

- The goal was not as high a priority as you thought.
- Your time management is not sufficiently effective.
- You have a psychological block about doing the work or achieving the goal.
- The goal is unrealistically high.
- You are not trying hard enough.
- Competing goals make the goal unattainable.

To better understand the source of the problem, first assess your situation. What is happening right now? On what issues are you deceiving yourself? On what activities are you spending your time? What goals are being met? Then evaluate your satisfaction with your present direction. What "course corrections" could you apply to improve your satisfaction? Next search for strategies that will bring you closer to these revised goals. Finally, structure your goals in a more detailed way. Flesh out the specific actions needed to make them happen.

If you feel that you may have deceived yourself about your goals, I suggest reading my book *Decision Power: How to Make Successful Decisions with Confidence* (see the Suggested Reading List), which can help you "reality test" your decisions. It offers guidelines for choosing goals that are aligned with your needs and personal values. *Decision Power* also shows you how to use affirmations to maintain enthusiasm and overcome obstacles that appear insurmountable.

To help you with the introspection needed to assess your goals, self-evaluation guidelines are given in Table 16.

TABLE 16. Self-Evaluation Guidelines

1. Am I meeting my financial goals in consulting?
2. Am I advancing professionally?
3. What are my emotional reactions to consulting?
4. Am I building a basis for continued success? How are my present activities laying the foundation for tomorrow's success?
5. Am I making good use of my nonbillable time?
6. Has my enthusiasm increased or decreased this past year?
7. What new areas should I be getting into? Am I planning time and resources to do this?
8. What "lessons" have I learned this year?
9. What aspects of my business need more attention?
10. Am I making good use of available capital?
11. Are my customers satisfied with my work?
12. Which of this year's goals remain unachieved? Do I still want to achieve them? What effort and resources would be required to achieve them?
13. How am I rewarding myself for good performance?

CHAPTER 16

The Technical Challenges of Consulting

Each technical specialty presents unique challenges, but all have certain issues in common. In this chapter, I discuss some of these common technical issues and show how successful consultants deal with them.

A MINI-ESSAY ON THE INFORMATION EXPLOSION

Your reaction to all the planning and development in Chapter 15 may be that it is too much work! Well, you are right. Your time is limited and unfolds linearly, while technology, which waits for no one, is expanding explosively (nonlinearly). This fact becomes more obvious as you gain professional experience.

As society's knowledge base expands to gargantuan proportions, individuals can no longer maintain significant domains of shared knowledge. That is, the amount of personal knowledge we share with others is decreasing. There is simply too much knowledge out there. (See Figure 17.) One result of this thinning effect is important to consultants. Technical specialists find it difficult to communicate with people in other specialties. They are unable to grasp the subtleties and jargon.

Consultants can no longer take for granted that clients will understand their analysis or reports, or the technical trade-offs in support of a conclusion. In the future, the burden of good communication will fall increasingly on the consultant. Success will come easily to those consultants who can explain their ideas and procedures in nontechnical language. This trend is similar to one taking place in the medical field. Doctors are being trained now to talk to patients in nonspecialist terms. Explaining complicated procedures in everyday language is extra work for the doctor, but it improves the doctor-patient relationship and helps the patient make more informed decisions.

The information explosion is accelerating faster than anyone's ability to keep up with it. How do you handle this as a consultant? Do you simply try

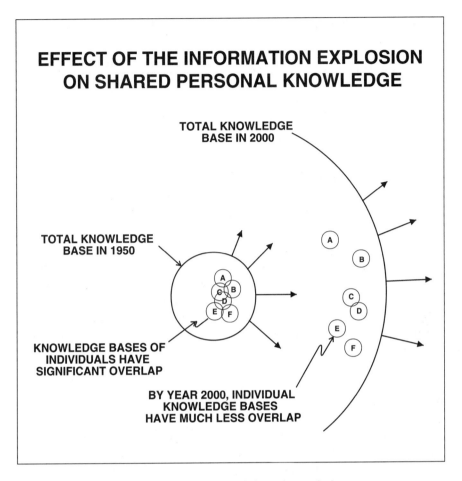

FIGURE 17. Effect of the information explosion.

harder to keep up-to-date? That's the answer many engineering societies will give you. Their councils on education stress the importance of keeping up-to-date. But ask yourself, "What *purpose* does staying current serve?" Just adding to your base of information is not likely to make you more creative or successful as a consultant. You can fill your head with the latest theories, data, and methods, but if you don't actually *use* them, such efforts won't advance you professionally.

That is, information has *value* that varies with the way in which it is used. You may be able to impress your colleagues with the story of how Le Verrier predicted the existence of the planet Neptune in 1846 by considering unusual perturbations in the orbit of Uranus. But unless you are working in a specialty that uses this knowledge, it will be of no benefit.

A qualitative description of the value of information is:

$$\text{VALUE OF INFORMATION} = \frac{(\text{USEFULNESS})(\text{NOVELTY})}{\text{COMPLEXITY}}$$

Besides usefulness, the other factors that affect the value of information are *complexity* and *novelty*. The more complex the information, the less valuable it will be. Complexity usually means that the information must be manipulated and processed before it can be employed in a beneficial way. Novelty, already discussed in Chapter 4 with regard to marketing technical services, makes information *more* valuable. By novelty, I mean that the information has not yet been widely circulated. If you are the only one who knows a bona fide stock market tip, you can make a killing on the market. If, however, you are getting your "tips" from a market newsletter that has a circulation of thousands, the information is proportionally less valuable. As time goes by, more and more people acquire the information, which dilutes its value. Therefore, don't waste precious time acquiring information you won't use right away! How does this apply to consulting, which is often *information intensive*?

BE A SURFER ON THE "WAVE OF KNOWLEDGE"

In slower-paced days, swimming through the sea of knowledge would get you to your destination. Now the winds of rapid change have picked up and the sea is no longer calm. The breakers prevent you from reaching your destination in the old-fashioned way. If you have ever tried to swim at a beach where the breakers were coming in, you know the feeling. The breakers come sweeping past you, faster than you can swim, making you feel small and helpless. In this age, you must learn to be a surfer on the waves of knowledge. How do you do this?

- The surfer is a competent swimmer in the first place. By analogy, you must have a solid grounding in the basics of your specialty. If not, you should go back and fill in the holes or review as needed.
- The surfer rides on the crest of the wave. She doesn't get bogged down swimming (plodding linearly) through the wave. To stay on top of the technology wave, spend at least one hour per week reading in your technical area. The emphasis here is on scanning. Read widely and quickly for general knowledge. Don't study or belabor the material unless there is a direct and immediate payoff.

- The surfer applies all of her effort to balance her board on a changing and developing wave. Likewise, you have to acquire a feel for how your technological area is developing: Which information-gathering efforts will, directly and immediately, keep you "up"? Teach yourself basics in new fields of interest that have short-term payoffs. I do this myself by devoting two hours per week to texts, journal articles, and short courses.

- Surfers don't try to surf in calm seas (dead technological areas). They use state-of-the-art boards (the latest learning technologies) to extend their performance.

- Surfers love surfing, but they don't surf twenty-four hours a day just because they own a great surfboard. It is OK to collect material in your specialty, but don't feel obligated to read it just because it sits on your shelf!

The analogy with surfing is not intended to imply that you should be superficial or follow fads. Fads have no rhyme or reason. Surfing follows mechanistic, repeatable laws that allow you to surf even though each wave is different. Once you ride the wave of information as a surfer, you will never feel the frustration of trying to stay up-to-date.

I purchase many books, journals, and software programs in my field. I use some of them from time to time, but I don't feel guilty about letting them sit on the shelf if there is no present application. Why do I collect information this way? If I use a technical reference for a useful method, equation, or explanation, I can save an hour or two compared with struggling with it myself or digging it out of a less appropriate source. Using a book just once in this manner justifies its purchase. Further, I find that owning a basic reference library in my specialty is essential to maintaining access to the data I need. Even though I live near excellent university libraries, these sources frequently do not have many of the basic books I need on their shelves. Just when I want them, they may be loaned out, discarded, or misplaced.

Keep information in its proper place—on your bookshelves. Using your head as a storage bin is very costly. Therefore, every time you are about to study an article, read a book, or take a seminar, ask, "What is the immediate payoff?"

Because the consulting business is information intensive, the ability to learn rapidly is all-important. Most fertile technical areas are completely new, so don't expect to find ready-made training courses to gain your competitive edge. Instead, define these areas for yourself and set about learning them as efficiently as possible. To learn efficiently, you should:

- Target your learning objectives precisely. Study only those topics that relate directly to your goals.
- Pick learning materials (books, articles, seminars) that are clearly and insightfully prepared.
- Move on to a new topic when you are stuck; don't shift down to first gear and try to slog through it.
- Practice! The more you practice self-directed study, the better you become at it. Commit to a program of lifetime learning. Set aside a few hours every week for professional development.

Mastery of new knowledge and skills is what gives you a competitive edge, or, as discussed in Chapter 6, your *unique selling proposition.* Once you have a mastery of these new areas, plan a path of *continuous* learning and improvement to *maintain* your competitive edge. Successful consultants are always adding to their "bag of tricks."

Consider the dilemma of Ed, a twenty-five-year-old consultant with a B.S. in computer science. He started his consulting business designing Web sites for businesses when he had only one year of professional experience and essentially no capital. His sole start-up expenditure was a computer and a $99 Web-authoring program. At first, Ed was one of the first "Web-heads" on the scene. Business was good. Two years later, however, Ed felt the pinch of intense competition from dozens of other consultants in his city who were offering the same services at lower prices. They had discovered that they could get into the same business with a minimum of equipment and one month's study to master the Web-authoring software. Moreover, in the two years since he had started, new software tools made Web-site creation simpler and technically less demanding. Ed's field was becoming too accessible: total beginners and entrepreneurial high-school students were flooding the marketplace!

As Web technology eventually becomes more accessible to Ed's clients, they will design their own Web sites. If Ed wants to stay in that field as a consultant, he'll have to pick new areas of Web design that are not standard fare and aggressively set about learning them and creating credentials in them.

BECOME EFFICIENT AT HANDLING INFORMATION

One of the most common complaints in the modern workplace is *information overload.* We are inundated with mail, interoffice memos, E-mail, "must-read" books, technical articles, brochures, and advertisements. As a consultant, you have an informa-

tion burden that will increase substantially, because you will be tracking a multitude of projects being pursued by a multitude of clients. To keep on top of your information, become more efficient in handling it:

1. Look at it only once. If it absolutely requires an immediate response, do it right away, but keep it very brief. If it can be thrown away, do it immediately. If it is important and must be saved, file it directly into its intended folder/bin. If you're not sure what to do with it, put it in your "holding" bin. After four weeks, if you don't recover it from the bin, throw it out!

2. Answer it only once a day. Use voice mail to collect your calls, and then call back when it's convenient for you.

3. Don't slog through the whole thing. I get *hundreds* of phone calls per year from salespeople who are expert at misrepresenting their intentions. When I ask, "Is this a sales call?" they usually respond, "No, I'm an adviser." Then they segue into their *grabber.* They claim that we talked three months ago and that I requested they call back now. Or they say I have won a free prize of $500, and I need only buy the new car they're selling (at full list price) to claim it. Some of these routines are extremely creative and engaging, but I have lost patience with them. If an unknown person calls and glibly starts what appears to be a sales pitch, I interrupt him and say he has ten seconds to state his business. If he can't, *click*! Similarly, when I receive mail marked "urgent" that appears deceptive, I give it ten seconds before I throw it away.

4. Find it with the "tab" system. When I'm reading a book or an article that I expect to refer to later, I place a small Post-It! note on pages of particular interest. Sometimes I'll write myself short reminder messages on the note, such as "Can I use this for the XYZ project?" This system has saved me countless hours in locating references. I can find the exact information I need without having to read the whole book again.

MEET THE PROS UP CLOSE AND PERSONAL

Rod Stephens is the president of Rocky Mountain Computer Consulting in Boulder, Colorado. He is also the author of three books, *Visual Basic Algorithms, Visual Basic Graphic Programming,* and *Advanced Visual Basic Techniques,* published by John Wiley & Sons. His educational background includes an undergraduate math degree from UCSD and graduate math work at M.I.T.. Soon after grad school, he started consulting; he has been at it for thirteen years.

Staying current technically has been an interesting challenge for Rod. "Some technical areas, such as the Internet, are moving so fast, one person can't possibly keep on top of it all." So, he spends three or four hours per week scanning Web sites and journals for general knowledge, but he doesn't study any single topic unless it becomes clear that there's a direct need for it. Once a month, he goes through his pile of magazines and throws out the junk issues, saving the ones that promise to be useful at a later date.

Rod says that his clients are also overloaded by the lightning speed of new technology developments, but that he has learned not to underestimate the client's intelligence. Rod is way ahead of the client in his domain, programming, but clients have different talents and perspectives that add materially to the successful completion of his projects. Clients know their primary knowledge domain very well, but most don't feel the need to learn Rod's programming domain at all. All they care about is whether he's giving them the right final results. On the other hand, Rod says that consultants must learn enough of their clients' domain to communicate effectively.

Over the years, Rod has learned a unique lesson about information management. "Staying on the absolute front edge of a technical development area doesn't pay off. On this 'bleeding edge,' technical issues can shift repeatedly before they stabilize. It's nerve-racking when you're trying to build something that still needs to work in six months. At the very least, you'll end up doing things over four or five times because of the incessant changes. To maximize my efficiency, I now keep my distance from the bleeding edge."

5. Think conveyor belt, not wine cellar. The old system of keeping reference materials postulated that information retained its value indefinitely. Information was like wine: the older, the better. You kept your vintage books, journals, and memos because they rarely became obsolete. In the fifties and sixties, this appeared to be true because technology moved so slowly. Now, to be efficient, your information-handling system must accommodate different rules of the game. No longer is your collection of reference material a *static* entity that must be kept forever. Your new system must be one that promotes a constant flow of materials. It's dynamic, like a conveyor belt. Therefore, date-stamp materials coming into your filing system. Make sure to have in *and* out bins. Periodically go through your collection and weed out obsolete material—don't let it accumulate.

ORGANIZE FOR MAXIMUM PRODUCTIVITY

Over the years, I have discovered myself solving the same problem for different customers. Of course, the particulars are always different, but the general methods are sometimes nearly identical. By changing a few parameters, I can reuse portions of a previous analysis. I believe this is a very common occurrence for technical specialists.

The software business has formalized this concept with the word *recycling*. Recycling takes advantage of the recurring nature of technical solutions and has proved to be an enormous time-saver and productivity enhancer. To make recycling work, always keep in mind that what you're doing today might serve as the basis for a similar project tomorrow. Therefore, take copious notes as you develop your design/analysis/solutions. Without carefully documenting your work, you may have to "reinvent the wheel" many times over. One or two years later, you will have forgotten the details. By placing comments in your work, you can figure out what you did and quickly retrace your steps. Furthermore, if you can manage it, try to *generalize* your processes so that they can be applied to other problems with minimal refitting.

Suppose, for example, that you are being asked again and again to solve the one-dimensional heat transfer equation for a composite slab. Create a general-purpose module that accepts specific input parameters for each particular case. You can pass the savings in time along to your customer if you are doing the project on a T & M basis. Or if you are charging on a per-job basis, this efficiency translates into higher profit for *you.*

The key to higher productivity in your consulting work is to systematize, to develop general procedures to handle common tasks. To continue the discussion, I must distinguish between five general types of technical consulting. Each type has its own parameters, its own "game," and its own challenge.

1. Guru work: This consists of high-level review of the client's practices, products, and strategic directions. The issues here are having the diagnostic skills to quickly assess the client's situation and the diplomacy skills to recommend changes in a nonthreatening way. The greatest challenge in this high-level work is that clients are usually looking for quick fixes or miracle shortcuts that will save them millions and reduce their risks.

Here are some strategies to be more productive and effective in this type of work: First, always have your suitcase ready. Leave 25 percent of your schedule open to be able to respond to these guru session requests, which usually last for a few days at most. Second, develop a set of "standards" for reviewing your client's practices, products, and strategies. Clients are more likely to accept your expert judgments when you can present a rationale for your assessments to them. Third, have a few "canned" presentations covering the most typical situations ready to go at any time. These presentations help educate your clients on the issues and create a basis for information gathering and detailed discussions. They show that you take your job seriously and are not simply trying to "wing it" by playing the all-wise but do-nothing mandarin consultant described in Chapter 3.

2. Project consulting: This is contracting to do an entire job at your own facility and producing a "deliverable" package as the result. The issues here are mustering the discipline and organization to work effectively in your own environment, communicating well to get necessary specifications and information, and finding a way to test the acceptability of your product and/or recommendations before it's too late to modify them.

To be more productive with this kind of work, create and follow a project schedule (described at the end of the chapter) to avoid embarrassing delays. Second, make sure your tools and suppliers "work" beforehand. Don't wait till the last minute to perform tasks that depend critically on untested components and sources of supply. Finally, create a work space with all materials at hand. Easy access to your materials and equipment will increase your productivity and make your work more enjoyable.

3. Technical gun for hire: This consists of supplementing the client's regular workforce in a specialized area, working mostly at their facility and under their direction. This issue here deals with image and control. You must figure out a way to project a "technical heavy" image without alienating the client's staff. Gunslingers in the Old West did this by showing off their fancy gun-twirling techniques. It stood to reason that anyone who could twirl his gun that well would be very fast on the draw. Likewise, you can employ similar methods given in Chapter 11 to enhance your image as a technical expert.

The issue of control requires that you not be too passive in accepting day-to-day directions from the client. Yes, they are paying you. Yes, you should be accommodating and respectful. But no, they don't *own* you. Make it a point to reserve one day a week for other clients or your own R & D efforts. It's financially tempting to work five days a week at their facility, but if you do, in two or three months they *will* think they own you!

4. Speedy technical service: In this kind of consulting, you render a service lasting a day or less, where you use your own expertise, equipment, or computer program to help a client with a well-defined task.

To become more productive in this line of consulting, develop standard methods for the client to more quickly specify the particular service they want. That is, create a *menu*. Also, set up an easy way for them to pay when the service or product is delivered, avoiding the hassle of billing them later. Obtaining a charge card account might expedite this. Finally, develop ways to track customers for a mailing list that can generate repeat business with special offers.

5. Expert tech support: This is offering your expertise on a time-and-materials basis where the time increments can be as small as fifteen minutes. Lawyers and computer support organizations do the same thing when clients call with short questions on an infrequent basis. They track the total time on the phone with you by jotting it down in their calendars or with a computer program that automatically monitors and logs the telephone conversations.

You can make yourself more efficient in this line of business by centrally locating your information in contact management or "support desk" software. When the client calls, you won't have to fumble around to find their file. You'll have all the information right at hand. Also, for this kind of work, prepare a standard service contract that explains how you will bill clients and track your time. It is not uncommon for this type of work to be done on a *retainer* basis. That is, the client signs a service contract for an overall amount that they prepay to you.

YOUR OWN R & D DEPARTMENT

Part of the technical challenge of consulting is creating *salable* technical skills. Developing these technical skills is not a matter of intellectual brilliance, but of simply taking the time to acquire specific knowledge and tools that will help persuade clients to give you consulting contracts. This personal research and development effort has three dimensions:

1. Developing competence. Developing competence is all about learning. In the middle of a contract, you won't have much time for this, so it must fall under your own R & D time. Ideally, the learning occurs on two levels. The first is the textbook level, where you read about a new subject and solve a few academic problems to make sure you grasp the content. In my estimation, this is passive learning. The second form of learning is more active and intense: it is creating real systems and working models where you use the subject material to actively synthesize a new product. As a rule of thumb, one ounce of active learning equals one pound of passive learning. There is no substitute for actually playing with the equipment, compiling that computer program, contacting the vendors—actually going through the footsteps of a real live project.

2. Developing tools. Like a mechanic putting together a personal tool kit, you will need to assemble your own set of tools, techniques, and data. With a well-developed tool kit, you will be able to offer a competitive advantage.

3. Forging technical contacts/relationships. In performing many technical contracts, whom you know—and who knows you—is often as important as what you know. Becoming familiar with the latest equipment, services, and vendors is a highly worthwhile R & D goal. Because vendor/supplier relationships take a long time to mature, aim at developing them gradually as you work in your field.

> Gene is a fellow consultant in heat transfer design. His ace in the hole is a commanding expertise in heat transfer *hardware*. He collects catalogs from hundreds of hardware vendors and stays on top of their latest offerings. Gene spends his R & D time learning how to use the latest products and getting the inside information on which products are reliable and which aren't. He makes it his business to know what every vendor in his specialty offers, whom at that vendor's company to contact, how much they will bend their price and service terms, and so on.
>
> I have seen this consulting whiz in action. When clients say, "I want to use Universal Corp.'s model X muffin fan for this application," his response might be "Gee, I wouldn't do that. They discontinued that model three months ago and replaced it with a plastic one that won't satisfy your specs. Why don't you consider Colossal Corp.'s new model Y, which came out last month? It meets all the specs and costs 40 percent less."
>
> You should see the look on the clients' faces! After they calibrate his advice and discover that he really does know all those vendors and application details, they treat him differently. "Say, Gene, please tell us what we need here. Thanks!"

In today's competitive marketplace, once you have a general competence, your next goal is to stay *one* step ahead of your competitors. That is, make sure you're developing skills that are truly cutting edge, skills that give you a competitive advantage.

> Two consultants are walking in a jungle. They are suddenly confronted by an angry tiger. The savvy consultant opens his briefcase and puts on a pair of running shoes. The other consultant exclaims, "Are you crazy? You can't outrun a tiger!" The savvy one says, "No, but I can outrun you!"

Tip

When you are breaking into a new technical area, you will spend a large amount of your own time developing technical skills. However, the busier you become with client work, the easier it becomes to shift some of the development costs to the client, in support of ongoing projects.

Economist Lester Thurow coined the phrase *zero-sum game* to describe this effect. In freely expanding markets, almost everyone wins. In saturated (competitive) markets, however, you gain market share only at someone else's expense. Depending on the market conditions you are facing, sometimes it's more important to figure out how to beat the competition than to solve your development problems from first principles.

PROBLEM-SOLVING SKILLS PAY OFF HANDSOMELY

Much consulting work consists of posing technical problems—and then finding the solutions. Hence, problem solving is an essential skill for consultants to master. Fifty percent of the technical challenge in a problem is learning how to *approach* the problem as a professional problem solver. To do this, hone your skills in decision making, information gathering, and critical thinking. Also, seek to "grasp the entire problem" instead of looking straight ahead in your narrowly defined scope.

The steps in the consultant's problem-solving process are:

1. State the goals—how the final result should "work."
2. Diagnose the problem.
3. Pose the problem in such a way that it can be solved.
4. Develop the criteria to evaluate alternatives.
5. Come up with alternatives.
6. Evaluate the alternatives according to the criteria.
7. Make trade-offs on problems that don't satisfy all criteria.
8. Orchestrate the decision process. Get others to participate in the process and to support the decisions made.
9. Justify the technical approach that was taken.

THE NATURE OF PROBLEM SOLVING

- Each problem is unique — if not in statement, then in context. This means that each problem poses a learning opportunity — if you are open to it. Learn to love learning! A delight in learning is the essence of the problem solver's personality.
- Realize that some problems are insoluble, some have a unique solution, and some have a multitude of solutions. Further, the resources and/or time required to solve a problem may not be available, forcing a compromise or "trade-off" solution where some goals are forsaken for the overall good.
- A truism among mathematicians is that a well-posed problem is already half solved. Therefore, strive to state the problem as succinctly and directly as you can. Sometimes this means reformulating it in different language or more generally, so as to increase the number of options available. Get to the heart of the problem — the objectives (what the client wants to accomplish) and the major parameters (what variables must be manipulated to devise a solution).
- Don't try to gather *all* the information you can about a problem before you start taking action, but only that which is absolutely necessary. Gerald Nadler and Shozo Hibino, in *Creative Solution Finding* (see the Suggested Reading List), call this the Limited Information Collection principle. Problem solving is not gathering encyclopedic information about the problem, but producing viable solutions. The data you gain from taking preliminary action are often the most important pieces of information you can obtain.
- Get working models up and running as quickly as possible. Explain their purpose to your clients and show how they provide benefit. Dummy up the parts of the project that you can't fully grasp at first. Make bounding cases, use back-of-the-envelope calculations, etc. The important thing is to model the behavior of the whole as fast as you can, so that you can get feedback from the client: Is this how you expect things to work?
- Get a firm idea of the client's goals with respect to the problem. What are they trying to accomplish? How does it fit in with the larger context of their business/position? What are the time and cost constraints? What would constitute an ideal solution to the problem?
- Always think of the next step. What happens after the problem is solved? Are there continuing needs that have not been determined or addressed? Does the current solution introduce new problems of

which the client may be unaware? Are recently breaking trends or influences from the outside likely to compromise the effectiveness of the solution you just produced? Have you provided a migration path for the client to accommodate future changes with modifications or upgrades? Build the solution to allow for future expansion, so your client will not be locked into a "design corner." A little foresight on your part gives the client greater value for his money—and gives you a steady stream of projects!

When you sharpen your problem-solving skills, clients will be very impressed with your ability to pull together the disparate aspects of the technical decision and "argue" the merits of the different options. I urge you to consider problem-solving meetings as opportunities to pull together an entire position. Don't just throw the facts on the table and wait for the client to structure the problem and evaluate the options. Prepare for the session. The more effort you exert in researching the criteria, options, and trade-offs, the more knowledgeable you will appear and the more credible your recommendations will be.

For example, on one project I was given a problem that turned out to have no solution within the constraints of the product's specifications. At first, I informally asked the project manager if one of the constraints could be relaxed. "No way!" he answered. "It's cast in concrete."

I spent two days preparing a detailed presentation, complete with calculations, showing why no solution was possible with the current configuration. Something had to "give" to allow a workable solution. The client bought into my presentation and asked me to suggest the lowest damage fix that I could

Tip

Many problems go unsolved—or are poorly solved—because the problem solvers are stumped in coming up with good alternatives. You, too, will discover that problem solving depends critically on the amount of *insight* and *creativity* you can bring to bear in generating alternatives. Toward this end, I refer you to *101 Creative Problem Solving Techniques: The Handbook of New Ideas for Business* by James M. Higgins (see the Suggested Reading List). In just 214 entertaining pages, it shows how to enhance your problem-solving ability with a number of creativity-stimulating techniques.

devise. I went back to my office and prepared a list of parameters, which, when changed, would allow a solution to be found. I made a new set of calculations that showed the advantages and disadvantages of each option. Armed with this material, we sat down and figured out which change would be the least painful in terms of schedule slippage, technical risk, and cost. By getting the client to increase a single parameter by 3 percent over specification, we found an acceptable solution.

MANAGE YOUR OWN PROJECTS

Congratulations! You got the contract—now you must do the job! Managing the technical project is more than just setting a work schedule, more than applying the problem-solving techniques I just discussed. Managing the technical project means developing a strategy to bring the technical part of your project to completion.

The *systems approach*[1] to technical management breaks the project into logically connected technical phases. In many cases, the technical work scope of your proposal becomes the basis for the chart of technical milestones. In other cases, however, you will be in a development project where the ultimate product and technical direction depend on a number of intermediate decisions that can only be made after work has commenced. These intermediate decisions are the result of feasibility studies and tests, evaluation of options with respect to available equipment and software choices, and other studies.

The technical management challenge is to identify these intermediate milestones and structure a decision-making session around each one that requires further refinement of the project's direction. On large projects, you may also be required to write a report at each milestone. The usual milestones are:

1. **Problem evaluation and/or working specification:** State the technical goals and set forth preliminary specifications for required parameters, functionality, and performance.
2. **Feasibility study:** Compare options in terms of their ability to satisfy the goals and requirements.
3. **Preliminary design:** Show all major features of the option that is chosen and sketch how the final product will look.
4. **Final design:** A comprehensive report on the features, construction, and performance of the final design, including demonstration of its acceptability.

---

[1] For more detailed discussion on this subject, see Thomas H. Athey's excellent book, *Systematic Systems Approach,* cited in the Suggested Reading List.

The case study in the appendix gives a complete example of how these phases dovetail in a real-life project. For the moment, the most important guideline to remember is: *Get the client's "team" to buy into your approach at every major step.* You will discover that you must "sell" the job twice: First, to get the contract. This is really just getting permission to help. Second —and more important—at every project milestone, you must sell the team on cooperating and supporting your approach. If you fail to recognize this second sales job, it will be difficult to obtain resources deemed necessary midstream in the project that were not explicitly mentioned in your proposal.

Once you have defined the project milestones, the next step is assembling the tasks into a time sequence that meets the project deadline. Creating a project timeline protects you from a common mistake: technical specialists (including technical consultants) often jump into the middle of a project without thinking it through. They impulsively start with the parts that are the most fun or the easiest to accomplish. In doing so, they often neglect the time lag that certain critical-path items entail. When the project deadline arrives, they are still waiting for parts or working on dependent tasks they should have started earlier in the schedule.

After breaking the job into bite-size chunks, identify the critical-path items in the project. In making the project schedule, sequence these critical-path tasks so that work proceeds without hang-ups or delays. Doing the project scheduling yourself ensures that the parts of your project are assembled in the right order at the right time. If your project is complicated or involves the efforts of many contributors, consider using project management software such as Microsoft Project to clarify task priorities.

One of the major considerations in technical management is structuring the project to minimize the risks. Especially when you are innovating or working in uncharted territory, you must acknowledge the risks and choose an appropriately safe course. That few things ever work the way they are first designed highlights the need to plan for contingencies and to fine-tune the system once it is "complete." In a hastily designed system, things have a way of going wrong at exactly the worst time, and one disaster often precipitates another. Murphy's Law and its corollaries state the situation perfectly:

MURPHY'S LAW: IF ANYTHING CAN GO WRONG, IT WILL.

Corollary 1: Everything goes wrong all at once.

Corollary 2: Left to themselves, things tend to go from bad to worse.

Corollary 3: It's impossible to make anything foolproof because fools are so ingenious.

The best way to avoid having problems is to face them head on, before they swell to unmanageable proportions. Identify critical issues and roadblocks early on. Give yourself the time and logistical elbow room to overcome new obstacles as they arise. Don't overload your client with a huge list of potential problems. You are supposed to be a problem solver, not a problem producer! Sift through and prioritize the technical problems. Keep the minor ones to yourself and bring up only the major ones with the client.

Murphy's third corollary is interesting because it implies that no matter how hard you work at anticipating the problems, it won't be enough. Trying to design a "foolproof" system is futile. When you design a system, people will sooner or later use it in ways that were not intended. Even a master carpenter will one day find himself in a pinch and drive a screw through a board with a hammer. Of course, do your best to anticipate problems and address them in the design stage. But don't be too sure that you can anticipate *all* problems. Allow sufficient time *after* the system has been designed to build in additional safeguards as required.

To minimize the technical risk in your consulting projects, I offer Harvey's Principle of Innovation, which is a takeoff on Occam's razor[2] in philosophy:

HARVEY'S PRINCIPLE OF INNOVATION

Be innovative, but to assure success without endless fine-tuning, apply the *least* amount of innovation necessary to bring the project to fruition.

The more new "knobs" (innovation factors) you add to your system, the more unlikely it is that you'll be able to make the darn thing work! When devising new systems for clients, resist the pressure to use all new or untested components. Use new parts sparingly — or you might not be able to fine-tune the system at all. If you must work with multiple innovation factors, integrate and test them one at a time. By breaking the innovation into smaller parts, you will be more able to cope with the fine-tuning process. Be sure to include time for "fine-tuning" in the work scope of your proposal to avoid alienating the client.

Another strategy to handle the risks of "innovation overload" is to create working models or develop test programs, similar to the "beta testing" that is so common in the software industry. Just do your best to assure that

---

[2] A rule in science and philosophy suggesting that the simplest theory explaining a new phenomenon is always preferable and that explanations for unknown phenomena should first be attempted in terms of what is already known.

TECHNICAL MANAGEMENT IN A NUTSHELL

- Gather necessary inputs, specifications, drawings, and materials.
- Digest the material for comprehension: Is it complete? Do parts contradict each other? Is it overspecified?
- Generate a solution strategy: Identify goals. Determine paths to accomplish them. Identify obstacles and develop ways to overcome them.
- Get the "team" to buy into your approach at every major step.
- Determine feasibility: Demonstrate "proof of principle." Create beta version or working model.
- Identify new challenges based on feedback and results of model.
- Implement solution, order parts, assemble final product, do final analysis.
- Present results, product, or analysis to client.

your working model is representative of the full-scale product. In creating fluid mechanics models, for example, dimensional similitude methods allow you to define the similarity parameters for any physical situation. By designing the similarity number to be the same for both model and full-scale, you ensure that a wind-tunnel model, for example, properly predicts the behavior of a full-scale aircraft. Not all technical problems can be cast into a mold this convenient, but I urge you to think critically about constructing useful *and relevant* models as a way of quickly zeroing in on your full-scale development.

CHAPTER 17

Decision Making

THE STORY OF GEORGE

The decision to become a full-time consultant is usually made under stressful conditions. Consider the case of George, a talented Ph.D. electrical engineer specializing in antenna design. He became disenchanted with his employer after being assigned to a two-year project that was certain to be dull and boring. He discussed the matter with his supervisor, but the company was pressed for experienced people,  and he could not be reassigned to a more interesting and challenging project. George decided to send out his résumé in search of other employment.

After a few months of interviewing, George found that the demand for his talents as a direct employee was lower than he had anticipated. Although there were five companies in the local area that could use his services, only one was interested right at the moment. They offered George a position with significantly less responsibility than he currently enjoyed. He also received an offer from a company on the other side of the country, but relocation was unacceptable to his family.

George became depressed about his situation. Each day at work was more annoying than the previous. George felt he was in a trap and started to question his self-worth.

IS CONSULTING A GOAL OR AN ESCAPE?

How can George make reasonable decisions under the above circumstances? First, George must examine his *situation* and *motives* as carefully as he can. Is consulting a goal or an escape from a negative situation? *Why* does he want to become a consultant? Is it the work itself? Is it the potential for higher income? Is it the independence of being his own boss?

The decision to go into full-time consulting requires a weighing, or trade-off, of your personal goals, your personal constraints and situation, and the business environment. The ultimate end of the decision process is to determine whether consulting is the best alternative among the many that are open to you.

Goals and plans were discussed at length in Chapters 12 and 15. It is important to understand that goals and plans are not decisions! Goals are desired states, and plans are proposed paths to reach those desired states. A decision, on the other hand, is a commitment to *a course of action* (or inaction, in some cases). All good works ultimately hinge on action rather than intention.

This chapter elaborates upon the decision-making process because many engineers have not had sufficient exposure to this important aspect of their lives. They are only vaguely aware of their motives and the emotional aspects of their decision making, having been educated to use purely rational and deductive processes in their engineering work.

Many engineers, like George, become disillusioned in their jobs and feel that they have been denied opportunity. It is easy for them to come to the often-heard conclusion that "engineering sucks." This complaint is not productive, however. Negative emotional energy drains you of essential physical and productive energies, leaving you without the power to change your situation. Denis Waitley's *The Psychology of Winning* offers a number of practical methods to increase personal happiness by rationally questioning these negative thoughts and attitudes.

In recent years, many people have asked, "I have been laid off and can't find suitable direct employment. Should I become a consultant?" The answer comes back to the issue of whether consulting is a goal or an escape. The tight employment market in corporate America does not make it easy to find a new direct position, no matter how qualified you are. Some people want to consult because they think it is an easy way out (that is, an *escape*). Yet, if you are not committed to consulting, it is a rough road to travel. People looking for an easy way out could better spend their energy by searching more vigorously for a direct position.

Laid-off (and retired) workers above the age of fifty with good skills and contacts have little to lose by going into consulting. In most situations, the chances of finding suitable direct employment at this age are significantly reduced. If they are willing to exert the effort, consulting may prove an excellent option.

Younger workers who have been laid off are a different story. A digression into consulting may detract from the kind of employment track record they want to build. I would encourage them to become consultants only if they are committed to it, have sufficient credentials and experience, and are sure that consulting fits in with their long-term career strategy.

HOW TO GET YOURSELF OUT OF A DECISION RUT

Looking back at the crossroads in our lives, we can see that we sometimes overreacted to situations because of our inability to deal with negative emotions. Too many of us operate near our emotional boiling points. When a major change occurs, we are too quick to label the event (or the people involved). In this manner, we prevent ourselves from looking at the situations as they really are. It is all too tempting to cop out by saying that the other party was entirely at fault and that your life is miserable because of those "bad guys."

A few years back, my friend Sally was discussing her career dilemma with me. She was distraught over recent rumors that her entire division might be laid off. She was a distinguished scientist but had no interest in marketing herself. In her view, her employer was the one responsible for marketing the scientists on the staff ("Scientists are not Fuller Brush salesmen"). Her function was to be the best scientist she could be.

Sally felt betrayed. She told me, "You know, life is much more complicated than we're led to believe when we're twelve years old." I agreed, but wondered to myself, "Yes, but who is twelve years old anymore?" The *emotional* content of this statement was transparent: "Help! I am confused and unable to sort out a clear course of action. The many conflicting goals and demands in my life are rending it asunder and depriving me of my wholeness."

In the midst of her temporary crisis at work, Sally lost all view of the fact that she still had choices. All she talked about was the anxiety she had over *not* having choices, about being constrained at work to march in place until the "final execution." In the course of our conversation, Sally's fears began to *concatenate:*

- fear of losing job
- fear of losing income
- fear of economic disaster
- fear of losing loved ones
- fear of being unlovable

I am not sure whether consulting would have been a good option for Sally, but I *do* know that she made her decision process much more stressful than it needed to be. Without mastering your fears, you remain blind to many good alternatives that lie before your very eyes. When your fears master you, you feel powerless to take *any* action. Fear of making a mistake holds you immobile and prevents you from trying the things you most need to do. Your self-esteem suffers because you are aware that other people can fend for themselves; why can't you figure your way out of this quandary? In short, you are in a "decision rut."

The way out of a decision rut is to realize that the penalty for taking no action is sometimes greater than the penalty for trying a new venture and

> **WITHOUT MASTERING YOUR FEARS, YOU REMAIN BLIND TO MANY GOOD ALTERNATIVES THAT LIE BEFORE YOUR VERY EYES.**

having it turn out badly. *Action is the key.* It doesn't make the risks go away, but it gives you enough forward momentum to overcome unsuccessful attempts. Just sitting there, confused and unwilling to move, leaves you with no momentum, no way out of the rut.

Negative emotional energies must be transformed into creative energies before a person can get out of a decision rut. The same energy that one person uses for complaint and despair, another will use to write technical papers, give seminars, create new software, and invent new products. You must learn to create, or negative emotions will torpedo your "ship of good intentions."

(This is not a pep talk! This is a reality you will discover for yourself. Enthusiastic people know that if opportunities won't come to you, you must go to them or create them.)

DOES IT HELP TO SEE A CAREER COUNSELOR?

My purpose in writing this book is to give you as much insight as I can about the technical consulting business. Yet, a book can address the issues only in a general sense. It may be useful to talk with a qualified career counselor who can discuss *your* particular situation. To clarify and expedite a decision-making meeting with your counselor, use Table 17, the Decision-Making Guide.

In discussing your decision, be sure to explore your attitude toward work. How much do you value professional independence and identity in contrast to security and corporate status? How much career and financial risk do you feel comfortable with? How strong is your urge to determine the particular projects you work on?

Career counselors tend to look at the broad picture in making career decisions. They structure the session in terms of your talents, inclinations, training, previous work history, and psychological disposition. They may offer testing (sometimes quick and superficial, sometimes sophisticated and expensive) to determine your ideal psychological work "profile." Because career counselors rarely understand the trade-offs or issues relevant to specific industries, don't ask them for tactical advice on specific opportunities or to review your business plan.

In some situations, you should seek advice from other sources. If your decision hinges more critically on evaluating the feasibility of your business plan or marketing strategy, look for the help of a retired business executive. As mentioned in Chapter 12, the SCORE (Service Corps of Retired Executives) service offered by the Small Business Administration is an excellent zero-cost source for this kind of advice.

Finally, if your decision depends more on issues such as risk taking and balancing priorities ("Dare I do this?" "Am I crazy to even consider this?"), I suggest reading Anthony Robbins's *Awaken the Giant Within* and taking a live business motivation seminar.

**DEFINITION: A DECISION IS A COMMITMENT
TO A COURSE OF ACTION.**

VALUES VS. RATIONAL CHOICES

Some people complain that weighing the pros and cons in decisions is like comparing apples and oranges, that it is not "fair." Such is life. Each option has its own advantages and disadvantages. It is up to each person to weigh them according to his or her own value system. In the field of career choices, there is no such thing as a *totally rational* choice. We are the creators of our own values.

Personal decision making is a combination of rational and emotional factors. Although it is important to base your decisions on objective facts and

TABLE 17. Decision-Making Guide

1. **State the problem.** What are you trying to decide? In the present context, it is if (and when) you should become a consultant. To make that decision effectively, you must state your goals, constraints, and priorities.

   - My goals are _____.
   - The constraints are _____.
   - My priorities are _____.

2. **State the options.** The business plan developed in Chapter 12 proposes one option—technical consulting—to solve the problem. Other options are:

   - Remain in your present position.
   - Move into management with your present company.
   - Seek direct employment with a new firm.
   - Start a commercial venture.
   - Consider how others in similar situations have solved this problem. But don't be limited by this; use your imagination.

3. **Determine how closely each option solves the problem.** Sometimes no solution exists within the given constraints. If this is the case, ask what constraints must be relaxed for a solution to become possible. Or ask how additional resources or support might be sought out to make a solution more feasible.

4. **Consider: Is the decision reversible? Does the decision have to be made now?** Sometimes the penalty for making an incorrect decision is small, either because the decision can be reversed or because there is a convenient fall-back position. Also, timing is often important, especially when you are in the middle of other demanding situations.

5. **Decide in a dispassionate moment.** If possible, never make weighty decisions when you are very high or very low. Emotional extremes, at either end of the spectrum, cause biases in perceiving the issues and evaluating the benefits and liabilities.

---

logical reasoning, not all decisions fit into categories that are amenable to logical analysis. Career decisions are not algebra problems! The latter have a unique solution that is independent of the person making the decision. Career decisions, on the other hand, involve *your* values, *your* feelings, and *your* situation. There is no uniquely "correct" answer. In choosing one alternative over another, you define yourself. Learn to feel comfortable making your own definitions instead of seeking "correct" answers that apply to other people.

ARE YOU SUITED FOR CONSULTING?

Table 18 is a consulting qualifications quiz to help you consider the suitability of your background and attitudes for technical consulting. Score the number of points indicated beside each question for a yes answer. For a no answer, score zero. Add the points for the entire quiz. If your score is above 200, your chances of success in consulting are very good. Scores below 100 warrant serious reservations about full-time consulting at this point in your career. Perhaps, with a few years of planning and effort, you will be in a better position to start your venture.

DO YOU NEED A PROFESSIONAL ENGINEER LICENSE?

In Table 18 the value of having a professional engineer license or other necessary license is assigned 15 points. For engineers, the issue is more complicated than a simple yes or no answer on this quiz. In many states, you must be registered as a professional engineer before you can *publicly* call yourself an engineer. Examples of public declaration are your brochure, business card, letterhead, and any advertisement or verbal claim you make about your services.

The need for being registered goes far beyond the issue of public declaration, though. If you *perform, in responsible charge,* an engineering service that affects, or can affect, the public welfare, then you are *practicing* engineering. Such practice requires, by state laws, a professional engineer (P.E.) license. The word practice includes engineering design, evaluation, and construction of projects, structures, equipment, and utilities for which compliance with specifications and bylaws is legally mandated. In particular, *certifying* an engineering design as adequate per applicable codes is the privilege solely of the registered P.E. Anyone who misrepresents himself in this capacity is liable to civil lawsuit and/or criminal proceedings. Our society has a legitimate right to demand that engineers performing such work meet standards of tested competence and professional ethics.

Having said that, it would seem that *all* engineers, direct employees as well as consultants, must be registered. Why isn't this the case in real life? In my estimate, only one engineer out of five is registered. At almost every large company, you will find many, many engineers doing "engineering" without a license. Are they breaking the law? The answer is no.

The explanation lies in the phrase "in responsible charge." In the eyes of the law, people who provide "engineering support" without being "in responsible charge" are *nonengineers*. Therefore, even though your thirty years of experience or your Ph.D. may constitute sufficient credentials to win and perform consulting jobs, if you don't have a P.E. license, don't offer your engineering work in a way that suggests that you are assuming responsible charge. It may hurt your vanity to disclaim your work in this manner, but it

TABLE 18. Consulting Qualifications Quiz

Question	Points	Me
1. Is my technical expertise marketable outside of my present position?	30	
2. Do I have an advanced degree in my specialty?	20	
3. Do I have at least ten years of professional experience?	20	
4. Do I have at least twenty years of professional experience?	10	
5. Do I have six months' living expenses in reserve?	20	
6. Can I work long hours without upsetting my family?	15	
7. Do I have a professional engineer license or other necessary licenses?	15	
8. Have I already done moonlighting consulting?	20	
9. Are there at least five potentially good customers in my local area?	15	
10. Have I ever worked (in a professional capacity) for a consulting or contract engineering company?	10	
11. Have I published four or more technical articles?	20	
12. Do I currently spend significant amounts of my own time improving my technical skills?	15	
13. Am I proficient at technical problem solving?	10	
14. Have I ever run a sole proprietorship or corporation?	10	
15. Have I written many technical proposals?	10	
16. Do I have sales or marketing experience?	10	
17. Have I been successful in group presentations?	10	
18. Do I find it easy to get along with all kinds of people?	10	
19. Can I tolerate occasional high-stress situations?	10	
20. Can I picture myself being happy as a consultant for the next twenty years?	20	
TOTAL	300	

is legally and ethically more appropriate. I see nothing wrong with telling your client that even though you are offering technically competent services, they must assume responsible charge for the work.

Because this book is aimed at an audience that includes scientists, computer software specialists, and other nonengineering technical specialists, not everyone needs a P.E. license. Ask consultants in your specific field if a P.E.

license is necessary. If it is, contact the Board of Registration of Registered Professional Engineers in your state and ask for a copy of the registration bylaws. For more details on the P.E. registration process, I recommend John D. Constance's *How to Become a Professional Engineer.*

I obtained my P.E. license after being out of school for nearly ten years. It took me thirty evenings of study to prepare for the required two-day written exam. In retrospect, this study provided me with a useful review of engineering basics. More important, though, being registered has made me feel more professional. It has given me a sense of identity with the engineering community at large and has made me aware that my actions are accountable to society.

DON'T WAIT FOR THINGS TO BE "PERFECT"

If your qualifications are adequate, and you really *want* to be a consultant, do it! Rely on your intelligence, intuition, and ingenuity to see yourself through once you have decided. Don't wait until things are perfect. It is certain that things will *never* be perfect. The motivational books in the Suggested Reading List can inspire you to overcome obstacles and trust in your own resourcefulness. Moreover, I can virtually guarantee that once you throw yourself fully into the situation of consultant, many good opportunities that you could not possibly have anticipated will arise. Trusting in your own resourcefulness and wanting the goal of professional independence with all your heart will allow you to survive the inevitable setbacks that occur in any long-term undertaking.

Nobody starts consulting with all the elements of success perfectly in place. Trying to do so keeps many capable consultants sitting on the sideline, waiting and waiting. My attitude is that many skills and resources are developed once you get started. As you go along, you'll learn by experience which elements need further development. Visualizing the set of attributes required for successful consulting may help you figure out the area(s) that need the most attention.

Figure 8 shows the six spokes that support the consultant's "wheel" and make it go around smoothly. If any of these spokes are missing, your wheel will "thud" every time that section touches ground. By looking at your own capabilities in relation to Figure 8, you can determine what your highest development priorities should be. At the center of the wheel is the prime requirement: a burning desire for professional independence. This passion will see you through the thuds and bumps, and propel you toward success as an independent consultant.

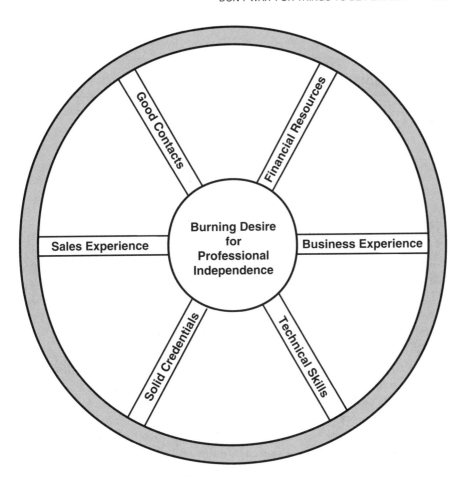

FIGURE 8. The wheel of consulting attributes.

CHAPTER 18

Making the Transition

WHERE DO YOU START?

If, after examining your goals and situation, you have decided to become a consultant, congratulations! Now, where do you start?

The nature of your transition from direct employee to independent consultant will depend on many factors that are particular to your situation. One especially important factor is whether you are planning to consult part-time or full-time. If you expect to do full-time consulting, you will need detailed plans to make an orderly transition. The idea is not to be obsessed with order, but to prevent a hang-up or delay that stops you from doing business or that cripples your cash flow.

The most important step, already discussed in Chapter 12, is to obtain moonlighting work or your first contract. If at all possible, this should be done while you are with your present employer. There are two good reasons for doing this. First, it will make you feel more comfortable in taking leave from the security blanket of your present employer. Second, people always ask if you are busy, and it helps your credibility to be able to talk about your first consulting project.

It is risky to make a transition to full-time consulting without a contract in hand. You would be better off suffering in your present situation until you have accomplished this important preliminary. In some cases, you may need to satisfy certain prerequisites such as building cash reserves or obtaining licenses. "Starting out" under these circumstances means addressing these preliminaries.

WHAT TO LOOK FOR IN YOUR FIRST CONTRACT

The title of this section takes a small poetic license: Instead of first contract (singular), you should be looking for first contracts (plural). When you start your own business, you must do something that you would *never* do as a direct employee: you must generate more potential work than you can actually perform. Yes, this means *proposing* more work than you can possibly do. Don't worry about what will happen if multiple clients want you at the same time. From your point of view, the worst thing that can occur is that you get to *choose* the project that is most beneficial to you.

Fishing is a good analogy for this situation. Just because you have a fish on the line does not mean that you will get it into the boat. Similarly, clients will tell you that they're *very* interested, that they're ready to cut you a contract. If you stop tending your other lines, you are a fool. In my experience, better than 50 percent of the time, these "sure bets" never materialize. The reasons are manifold:

- The client may have found a better way to solve the problem in the meantime.
- The client's management may veto your involvement on the project for reasons of cost, technical background, or schedule.
- The problem may "go away," i.e., turn out to be a nonproblem.
- The client may be overwhelmed with a panic problem that forces him to shelve your project for an indefinite period.
- The client's top management may issue a sudden cost-cutting imperative to "fatten the goose" for the quarterly earnings report.
- The client may not be able to make up his mind.

The client may tell you that your contract is a "done deal," that you should put off all your other projects and reserve next week for him. But listen to

 ## Worth Its Weight in Gold

Having more than one job going at a time makes you more efficient. You quickly develop the ability to work smarter and more effectively. You will feel more independent and powerful because you're not depending on a single client to provide all your business. Your revenue flow will be higher. Yes, you may work more than forty hours in a given week, but you will work more happily and with less stress.

your common sense. Has he backed this up with a written purchase order? Has he finalized the date he wants you to commence work?

Moreover, even when you land that big fish, there's no guarantee that it will stay in the boat. Every experienced fisherman can tell you a story about the one that jumped out of the boat *after* it had been landed! Similarly, experienced consultants can tell you about contracts that fizzled after they had been promised, after they had been officially signed, and even after work had commenced! There are no guarantees about individual fish. As a fisherman, the only thing you can do is *plan* for a certain percentage of them to never make it into your frying pan. That some get away is not a deterrent from an activity where a reasonably high percentage *are* landed and enjoyed.

In other chapters, I have already discussed how deadly it is to work five days a week at a particular client's office. If you do, the client will *know* that you do not have an abundance of other business. When the client senses that you have few other options, they will figure that you need them more than they need you — and treat you accordingly. Many clients may insist on your presence five days a week, but resist with all your strength and ingenuity. Always try to negotiate four days per week maximum, and only for a limited number of months.

The only way out of this trap is to *always* fish with more than one line, to *always* have more than one fish on the line. You will never be tied to one client, and you will always be planning and choosing the work that is best for you. The only way to retain a strategic advantage is to not need any individual client so badly.

What kinds of contracts are the best for beginning consultants? I strongly recommend that you select initial contracts that are *clearly* not in conflict of interest with your present employer, even if it means taking a project that is outside of your main interest area or at a lower billing rate. It is difficult to have the self-control to do this, because the companies that show the most interest in your services may be your present employer's competitors. Avoid anything that is shabby or unethical. You are creating a track record, and your reputation will follow you around!

How large should your first contract be to satisfy your need for "something" in hand? Is a contract for three hours of consulting sufficient for the intended purposes? Clearly not. You will need a contract for at least a few months' work before the rewards balance the risks. Contracts smaller than this are generally not enticing enough to precipitate a change, unless your particular form of consulting involves *many* customers, each of whom utilizes your specific service for a short time.

MEET THE PROS UP CLOSE AND PERSONAL

Leonard Schwab has been a consultant in electrical engineering for fifteen years. After receiving a doctorate in this subject, he went to work for several systems engineering companies in the Washington, D.C. area. He joined Lincoln Laboratory in 1979. His first task was to predict how local rainfall affected satellite link availability. At the time, the Defense Department was able to predict link availability for a single location but could not handle the technical challenge of all worldwide locations. Len developed advanced mathematical algorithms based on the Crane worldwide rain model and developed a unique method to plot the availability for any location in the world on a Mercator map. Within three years, Len published three IEEE Symposia papers that established his expertise in this field.

In 1983 a large commercial company doing development in this area heard about Len's research and started to pursue him. They needed someone with his exact background to work on a high-priority project. The technical manager on the project took Len out to lunch at a fancy restaurant and asked, "Would you be interested in switching jobs for a 20 percent salary increase?" Len was flattered, but he had been thinking about starting his own consulting company for some time. He told the manager, "Thanks, but the only way I'll work for you is as a consultant." The manager thought for a moment and shook his head.

The manager called Len once a month for the next three months to see if there was some way Len would reconsider. Len held firm. Of course, it is fair to assume that the manager was trying to hire another expert during this period. The manager had no luck in his efforts, so the next month, he called Len and agreed to give him a consulting contract. Len negotiated a 1,500-hour contract at $55 per hour (which was an excellent rate in 1983). With this first contract signed and delivered, Len felt confident enough to quit his position at Lincoln Lab and start his own consulting company. In the course of the next two years, Len made many new contacts at the client's facility that resulted in lucrative follow-on work and a solid foundation to expand his client base.

LEAVING YOUR PRESENT EMPLOYER

One of the considerations that bother many starting consultants is how and when to leave your present employer. You are probably in the middle of many projects needing your specific talents and contributions. The timing and manner of your departure will affect your chances of obtaining consulting work from your present employer as well as a good recommendation.

Strive to leave your employer on good terms. This will mean resisting the impulse to tell certain individuals what you really think of them. In giving your notice, try to allow enough time to put your projects in a form in which they can be taken over by others. Do not worry about how or when your projects are going to be *finished.*

Two weeks' termination notice is considered a minimum in the engineering business, and a month is typical for a situation in which relations are cordial. If your company has a reputation for pushing people out the door once they give notice, take this into account in your plans.

If you stand in good graces with your present employer, they may want your consulting services after you leave. However, do not *assume* that they will give you a fat consulting contract the moment you step out the door! In fact, the psychology of the situation works against you. Your former colleagues may envy your step to independence and react with negative feelings. They may feel that you have betrayed the company — and them — by leaving. If you are "right" in moving on, the others may feel "wrong" in staying.

WHY SOME THINGS SHOULD REMAIN SECRET

Because of the resentments just mentioned, some of your "friends" at your present company may change face the moment you announce your departure. In starting out, therefore, *do not discuss your moonlighting or plans* with associates from your present employer or with people in close contact with them. Your intentions will seem like disloyalty to these associates, and you may be forced to make a transition earlier than planned. Further, your associates' advice and reactions will be biased by their own interests in the situation. It may take great restraint to refrain from talking about your plans, but find a way to manage. It's to your great advantage.

While communication of your intentions is part of family sharing and consideration, be sure not to overburden your spouse, children, and loved ones with the details of your plans in making your transition. There are a couple of good reasons I make this recommendation. First, excessive talk about plans can generate resistance to your new activity. Your family may fear that

- you won't have time or money for them;
- you might fail and humiliate them;
- you will outgrow them;
- your life will become more exciting than theirs;
- your uncertain income may negatively impact their lifestyle.

Second, once your plans are set, *action* is more important than words. Discussing your intentions with everyone around you is not going to produce results. Spend the time in useful activities such as writing technical papers, marketing, or designing your brochure.

I can testify that the impulse to talk excessively about one's plans is very strong when starting to consult. Looking back at those moments, I can see that the energy I spent in talking could have been better used in doing the very things I was talking about!

WHERE TO GET ADVICE

If you need advice, get it from knowledgeable people who have nothing to gain or lose from your actions. In practice, this means *paying* for the advice. The danger of going to someone who has a stake in your actions is that you get to hear what is good for *him,* not necessarily what is good for *you.* Listen only to those advisers who can discuss the risks and rewards from *your* point of view. Also, don't go to a failure or a cynic for counsel. You will only hear about the dangers of enterprise and the futility of exerting yourself.

> John was a twenty-seven-year-old engineer at a large firm that built municipal waste incineration systems. John's specialty was combustor design. For the past four years, he had overseen the efforts of Herman, the consultant responsible for doing the designs. At their last meeting, Herman mentioned that he was discontinuing consulting for the next two years; he had been offered a visiting professorship at the university and would be overcommitted timewise.
>
> John saw this as a golden opportunity to become a consultant himself. After all, he was the only other person who was intimately familiar with the combustor systems. Here was his chance to get paid twice what he was currently earning!
>
> John wasn't sure, though. He had doubts about his ability and credentials. In private, he asked a co-worker and fellow engineer, Mary, what she thought about the possibility.

Mary said, "Why, that's a great idea. Nobody knows these systems better than you, John. Even Herman slipped up last year on the Middletown project, and it was you who caught his technical error. I'm sure you could get a high consulting rate from the company; after all, you schmooze with the folks in purchasing."

It turned out that Mary wanted John's job. Encouraging him to leave the company and become a consultant was a difficult choice for Mary because John was a friend. In Mary's book, though, career advancement came before friendship. She told John, "Go for it" and fed him *partial* evidence that supported his inclinations.

Although John knew combustor design well, he did not have sufficient credentials or experience to market his services to clients other than his present employer. Further, he was not aware that Herman was certifying the finished designs with his P.E. stamp. John's employer continued to use Herman over the years because it was an economical way to get their designs certified. John did not have enough experience to get a P.E. license in the next two years; he had not even begun to take the prerequisite licensing exams.

A month later, John was about to quit his job and announce his consulting practice when he happened to talk to Herman. His assessment differed from Mary's. Herman told John, "It would be risky for you to pursue consulting immediately. Why don't you gather the credentials you need first and then give it a try?"

In getting advice, get more than one viewpoint. Because many technical consultants are so immersed in their activity, they may not be able to offer you the best advice. They can't see the forest for the trees. Seek out other points of view—from accountants, businesspeople, and individuals who are knowledgeable about your particular markets.

Finally, just as there is danger in not getting enough advice, there is also danger in asking for too much. At some point, you can overdo it. Too many advisers are like too many cooks; they spoil the soup. When you reach the point of diminishing returns, feedback from your actions will provide better information than advice from others.

SELECTING YOUR COMPANY NAME

Select a company name right after you make the decision to go into business for yourself. You will be ordering stationery and brochures with your company's name on them, and it is essential that you keep these consistent.

Although names such as Xymox, Ajax, and Continental Supreme Enterprises may be appropriate for manufacturers of tangible goods, they do not reflect the personal service nature of consulting. Your personal name and credibility are strong selling points in this business and should be considered in creating your company name. Typical consulting company names are

- John Smith & Company
- John Smith Associates
- Smith Research
- Smith Software
- Smith Laboratories

Of course, you can always do business with simply your own name. Many famous consultants do exactly that. However, in many instances, the "Associates" or "& Company" highlights the fact that you consider yourself a separate business entity and not fair game for your client's employment recruiting office.[1]

A less preferred name is John Smith Enterprises. This has a flavor of commercialism that detracts from the professional image you are trying to project.

If your company consists of a *group* of consultants all in the same technical specialty, the company name could reflect this strength:

- New England Seismic Engineering
- Boston Noise Control Associates

However, make sure your company name is not so narrow in scope that it limits future business expansion!

If your name is exceptionally hard to pronounce or spell, use a company name that is an abbreviation or phonetic simplification. For example, a good company name for <u>W</u>olfgang <u>E</u>. <u>L</u>angsamschnitter might be WEL Associates.

Choose your company name carefully at the outset. It is not a good idea to change it frequently.

YOUR TIMETABLE TO SUCCESS

To make an orderly transition, develop a timetable including important factors such as the beginning of your first contract, the amount of notice you must give to your present employer, your cash flow situation, and the delays for obtaining needed permits, equipment, and office space.

---

[1] In the eyes of the law, a sole proprietor is an individual regardless of the company name. Thus, in legal agreements and contracts, you may be identified as "John Smith, dba John Smith Associates" (*dba* stands for "doing business as").

TABLE 19. Typical Transition Timetable

Time	Task
- 8 Months	Write a business plan that describes in detail how you are going to support yourself for the first year of operation (see Chapter 12).
- 6 Months	Get your first moonlighting contract(s).
- 5 Months	Decide whether to proceed with full-time consulting. If yes, then decide on a company name.
- 4 Months	Create your strategic marketing plan, as described in Chapter 6.
- 4 Months	Obtain local business licenses, if required. In many localities an individual doing business under his own name, e.g., John Smith Associates, is not required to have a license. For classified government work, you may also need a company security clearance.
- 3 Months	Start a company checking account.
- 2 Months	Determine the location of your office and make arrangements for furniture, equipment, and lease (if you are renting).
- 1 Month	Print business cards, résumé, brochure, and stationery.
- 2 Weeks	Arrange for a second (business) phone line, if required.
Time = 0	Give notice to your present employer.
+ 1 Day	Call your professional contacts and inform them of your new situation. Lay the groundwork for visiting them with your new brochure.
+ 1 Week	Obtain insurance coverage. Your present employer may be providing your health and life insurance. You will need your own coverage once you terminate employment. You may need business insurance as well. (See Chapter 13.)
+ 2 Weeks	Leave your employer. Proceed full-speed ahead in implementing your marketing plan.
+ 2 Weeks	Set up a schedule that reflects your new priorities as a consultant. If you don't manage your time, you will find yourself continually falling short of your goals.
+ 3 Weeks	Inform all professional societies, technical committees, magazines, etc., of your new address.
+ 3 Weeks	Submit a news brief to industry journals in your technical field. Most industry journals have a "People and Events" column that is ideal for this one-shot free advertising. The typical format for such announcements is: "John Smith announces the establishment of John Smith Associates, a consulting firm specializing in noise reduction for power plant applications. A former manager of ABC Noise Control Corp., he is an internationally known expert in.... Offices are located at...." Include a recent photograph of yourself with the news brief.
+ 1 Month	Call the IRS for quarterly estimated tax forms (and SS-4 if applicable).
+ 2 Months	Set up your accounting system.

A basic transition checklist is given in Table 19, where Time = 0 refers to D Day, the day you give notice to your present employer. I have chosen this as the reference point because it marks your "zone of irreversibility." Before D Day, most of your transition preparations are *reversible*. That is, you can decide to *not* go through with your consulting business without losing too much credibility. After time zero, however, you have openly declared yourself. Reverting back to your old situation may be possible, but not without losing significant credibility.

TIME MANAGEMENT

Time management is critical in making a successful transition. Consider that you have suddenly been promoted to president of your own company. As such, your attention must now be spread over a wider variety of topics than was previously required. Balancing the resulting work and information load can be overwhelming unless you consciously apply the principles of effective time management. I recommend Jeffrey Mayer's short book (see the Suggested Reading List) to help develop a system that works for you. For now I offer the following guidelines:

1. Every week, make a master list of all important tasks that you plan to handle. (I use a spreadsheet for this purpose.) Just enter the items as they come to mind. At this point, keep the descriptions short and don't bother to figure out which are the most important.

2. Take your master list and prioritize each task according to its importance. In my system, I give "must-do" items a priority of 1, "high-priority" items a priority of 2, "should-do" items a priority of 3, "do if time allows" items a priority of 4, and "can wait" items a priority of 5. (In my spreadsheet, after entering the priorities, I do a sort on that column to get an ordered list of priorities. I update this spreadsheet every week. Items that are completed are deleted, new items are added, and remaining items are reevaluated for priority.)

3. Using your prioritized task list, enter individual tasks into your daily planner. (Your daily planner is a critical component of your time-management system; you should always carry it with you.) In entering tasks, assign your highest priority tasks to your highest productivity

hours. If you are a "morning person," for example, assign your most important task to the morning hours. Use the other hours in the day for lower priorities, opening your mail, studying technical articles, and other less demanding chores. Also, if you are dealing with a long, complicated task, divide it into small pieces that can be accomplished in a one-day time frame.

4. As a habit, start each workday by looking at your daily planner. Make sure you understand what your number one, two, and three items are.

5. At the end of every day, look at your daily planner. Cross off each task that you completed that day. Most important, enter/revise *the next day's* top three priorities, in order of importance.

Keeping track of your daily schedule takes a few minutes every day, but it pays handsome dividends by saving you *hours* of wasted effort. It also gives you more awareness of how well you are accomplishing your goals on a daily basis. If a particular task seems to elude completion, break it down into smaller components. If it still eludes you, ask yourself why this is happening: Are you lacking certain materials or preliminaries? Are you procrastinating? Are you afraid of starting it? Is the task still a high priority? What would it take to get you moving on it?

Finally, using your daily planner, you can save a significant amount of time by combining errands. For example, I always try to combine trips to the bank, post office, and stationery supply store. Or, when I meet an associate for lunch, I combine that with any errands that would take me near that locale.

<u>Warning</u>

After you have been away from your office at a client's facility for a month, it's easy to fall into a psychological trap: When you're at the client's facility, you're "working," and when you're at your own office, you're "off." This attitude can lead to problems. The solution is to train yourself to think of time spent at your own office as "work" time. I know some consultants who dress up every day as if they are going to visit a client, even when they are working at home.

Recently, I finished a project that was done at the client's facility. It was a high-pressure project in which the client had been trying to micromanage me. At the end of every day, I was asked to give a detailed report of the day's progress as well as a list of the next day's objectives. The day after I finished this project, I just wanted to goof off. I took the clock off the wall, settled down in a comfortable chair, read a few magazines, and daydreamed. It was wonderful to spend a day without glancing at the clock!

This is fine—as long as you don't make a *habit* of the practice. It's great to take a mental-health day after an intense and stressful project. Just don't equate being at your home office with being on vacation. To be successful, you must be very productive in your own office. Don't tempt yourself by taking an *extended* vacation to let off steam.

AT HOME ≠ ON VACATION

In fact, one of the best times to take an away-from-home vacation is just after an intense project. Being away from your usual surroundings allows you to gain some psychological distance from the project. You will return to your office refreshed and recharged.

CHAPTER 19

Playing Your Game

CONSULTING IS LIKE A GAME OF TENNIS

In tennis, the stronger and craftier player forces the opponent to play "his" game. The weaker player always seems to be *reacting* to the moves initiated by the stronger. Her weaker knowledge of tennis—its tactics and strategies—and possibly weaker knowledge of herself contribute to a feeling of being off balance.

Consulting is like tennis in that you always have a person on the other side of the net—the client—who engages you in a "game" of sorts. Although the score is not as easily tallied in consulting, and although each rally does not result in a definite point gain or loss, this tennis analogy is useful in understanding the finer points of the consultant-client relationship.

THE "SECRET" TO PLAYING YOUR GAME

It is not easy for the consultant to get to the point where she is playing "her" game. There are no secrets or tricks to *playing your game:* you must indeed be the more knowledgeable and capable party! Further, you must be the one who is more in demand. That is, the client must need you more than you need the client.

In establishing a consulting business with no experience and little capital, it is inevitable that you will play your client's game for a while. By this I do not mean that your customers will necessarily intimidate you. But you will need them more than they need you, and the terms and tone of your interaction will reflect this fact. Extra effort will be required to adapt to their style.

Starting out, you have little credibility to suggest that *your* standard practices are reasonable or desirable alternatives. The client may insist on *his* procurement terms, *his* method of technical interface and direction, and the use of *his* equipment and software.

WHAT "PLAYING YOUR GAME" MEANS IN CONSULTING

As you gain credibility and a customer base, you will be able to pick the projects and clients that are most closely aligned with your own interests. When (and if) customers realize that you are a valuable and scarce commodity, they will be more agreeable to doing things *your* way. Some aspects of "your way" may include:

- Having sufficient leeway to use your methods and equipment.
- Having the customer send purchase orders in a timely fashion. Some clients reserve your time in advance, delay the purchase order, and then back down at the last moment. Without telling you, they have "parallel pathed" their options for getting the consulting work done. That is, they simultaneously explore many ways to solve the problem: Consultant A, Consultant B, and offer of direct employment to Person C. When you develop good rapport with your clients, they will not lead you down such a path without informing you.
- Having enough backlogged and imminent business so that a single dropped or delayed contract doesn't leave you with a block of unbillable time.
- Being able to do the work at your own offices and setting your own hours.
- Having nonbillable tasks identified and planned in advance, so as to fill gaps with useful activities.

SELF-KNOWLEDGE IMPROVES YOUR STRATEGIC ADVANTAGE

Another aspect of playing "your" game is finding the most effective use of your time and effort. As you start out, you will not have a large database for deciding how to prioritize your efforts. Experienced tennis players attain greater competence and self-knowledge by playing against a large number of opponents and observing their own reactions. After a while, they learn to pace themselves, to lead into their own strengths rather than to expose their weaknesses, and to read the signals that opponents unconsciously telegraph. Likewise, it shouldn't take you too long to learn

- which clients and market areas bring you closer to your goals;
- how to plan for new kinds of business;
- how much relaxation, exercise, and family time you need to remain "human";
- how to turn a client's casual inquiry into a paying contract;
- when to pay others for mundane tasks that do not utilize your best abilities;
- which technical conferences are worth attending;
- how to set daily priorities and monitor their achievement;
- when *not* to respond to certain requests for proposal.

The last item deserves more discussion: Playing your game means that you can refuse to play certain opponents when you sense their game is "tilted." Being able to say no helps shield you from situations where you are at a predesigned disadvantage.

In my experience, these no-win situations have certain common elements and identifiable patterns that allow them to be predicted. When three or more of the following elements appear, you are skating on thin ice:

1. They approach you, and you have never heard of them.
2. They insist that the job will take you only a few days, without giving you the chance to examine the scope more carefully.
3. They want it done ASAP.
4. They are reluctant to write a purchase order for the job.
5. They explain that you'll get paid when they do.
6. They have difficulty defining the job to you and describing its bounding parameters.

STRESS REDUCTION FOR THE ADVANCED PLAYER

In my opinion, not being able to handle the stress of dealing with clients is the number two reason consultants go out of business. (If you read Chapter 5 carefully, you already know number one: not being able to get enough business.) Yet, the psychological hazards of stress and its countermeasures are rarely discussed, even among veteran consultants.

This section is a "personal therapy session," a crash course on emotional survival for consultants. My goal in presenting this material is to give you some of the tools and insight you'll need to handle your *feelings* in the consulting process.

Consulting, like any other "helping" profession, deals with assisting others achieve *their* objectives. Psychologists and others in the helping professions have long noticed that being so intensely "other-oriented" makes helpers emotionally vulnerable. Your clients may be satisfied with the responsive way you have addressed their needs, but many of your own needs may be neglected or suppressed. The constant psychological stress of focusing so exclusively on the client's problems leads to burnout unless you can find a way to help *yourself* in the process. To survive in the consulting profession, you must consciously address your own needs and feelings.

The money you earn helping others provides you with income that may be very gratifyng, but income alone will not immunize you against burnout. You must create a *program* of continuous renewal to overcome the psychological pressures of consulting. Here is my personal program, which I have divided into three domains:

1. Set appropriate boundaries.
2. Acknowledge the bruises-and move on.
3. Nurture self-esteem.

Set Appropriate Boundaries

When you and a client start working together, neither of you has the other calibrated. During this honeymoon phase, both parties seek to learn how much they can ask for and how much they are expected to give back in return. The beginning of this discovery process feels like an awkward dance. Both the client and you will yo-yo until you learn to dance the same steps together. Eventually, both parties come to an "equilibrium position" where the pushes and shoves from both sides are balanced.

The equilibrium position assumes many dimensions. For example, in the dimension of trust, you find out that a particular client may be too trusting or not trusting enough. Eventually, you both come to understand where the other's boundaries are.

When a clinical therapist sees clients, one of the major concerns is setting appropriate boundaries for the interaction. Even though the client is paying, the therapist is the one with the lion's share of the control over the direction the therapy takes. In fact, clinical therapists undergo extensive training to be able to guide the interaction in acceptable directions. Even though consulting is a different "game" than therapy, consultants must strive to set similarly appropriate boundaries. For the consultant, this means:

- Refusing to let the client micromanage you.
- Insisting that schedule deadlines be negotiated, not dictated.
- Rebuffing efforts by the client to treat you like a direct employee. The client may forget that he is your *customer,* not your boss. Sometimes he must be reminded of this fact.
- Blocking attempts by the client to take unfair advantage of you or compromise your integrity. Set *moral boundaries* — limits to what you will and won't do. For example, a client asks you to falsify your test results "so the project can progress without hang-ups." If you go along with the request, the client may be happy, but *you* will be in trouble.
- Not meddling in the client's business. Never reprimand a client or try to improve them, unless you are specifically asked to do so.
- Making sure that *you* don't compromise the client's interests. (See my list of NEVERs in Chapter 10.)

You can't manage the *client's* feelings. All you can do is give them good service and hope they react positively. You *can* manage your own feelings, however. Toward this end, I would like to borrow an analogy from *Star Trek.* The consultant's "prime directive" is: *Carry yourself with dignity, no matter what.* Whenever a client gives you flak, resist or sidestep as the situation demands, but *always* carry yourself with dignity. Maintain your sense of professionalism in all consulting activities. Without it, you expose your reputation to grave danger.

Acknowledge the Bruises—and Move On

Sometimes an individual on the client's staff may insist on playing hardball. He doesn't do anything illegal, but like a second baseman who is blocking the base, he yields no room for you to approach without someone getting hurt. The only alternatives are to accept the out or fly into second base cleats first. There are risks associated with each alternative. With the first, you accede to bullying behavior of the client that will doubtless hurt your reputation. With the second, you fight back. Half the time, the client will respect you for it and cut you more slack. The other half of the time, the client will resent you for it and lie in wait to get revenge at the first opportune moment.

Whichever way the situation turns out, both sides wind up with bad feelings. Clients tend to be less bothered by their bad feelings, however, because they are in a better position to spin the account of the incident in their favor.

The consultant has a more difficult time learning to deal with these negative experiences. She may be able to accept the bruises on an intellectual level, but working through them on an *emotional* level is not as easy. And emotions are not easily dismissed.

So what to do? You can't undo the past. What's done is done. The important thing is not to let a bad experience throw you off track. Accept that many consulting relationships won't work—can *never* work—by virtue of who you are and who the client is. Don't go into a funk where you do nothing for a month except feel bad for yourself. Instead, reaffirm the validity of your goals and ask:

- What have I learned from the experience?
- What could I do differently the next time?
- Were there any warning signals I should have heeded?
- Is there any way to salvage the relationship? Do I even want to do this?
- What additional "people skills" would help me better handle such a situation in the future?
- Am I repressing latent anger about the way I was treated?

Not all problems are your fault. Maybe *you* should be angry. If you bypass justifiable anger at being hurt, it will doubtless find a way to turn inward. Under such circumstances, anger is the first step in the psychological process of healing and putting the problem behind you. Many technical people don't know how to process their anger properly. Instead of dealing with it or talking it out with friends, they feel guilty for the failure and blame themselves unfairly.

It's very important to openly acknowledge your wounds and let the healing process take its course. But I will tell you a little secret: Lasting peace of mind comes only from proving to everyone—and yourself—that the bad experience was a fluke. The only way to do this is to pick yourself up and get another contract in which you are victorious and successful.

**The best way to get over a bad experience
is to create a string of good ones.**

Nurture Self-Esteem

If you are going to prevail and not merely survive, you will need a healthy measure of self-esteem to envision new goals that capture your imagination and fire your passion. Abundant self-esteem is also your most important asset in handling the psychological pressures of dealing with clients. Learn to cultivate your own self-esteem, because *no one else* will do it for you! It is unrealistic to expect a client to say things like:

- Wow, you're very talented in what you do.
- You handled that technical problem very well.
- Congratulations on being so resourceful.
- Gee, you're a wonderful person!

Although most clients will never say any of this, you still need to hear it from someone to feel good about yourself. And that someone is you! The section on motivation later in this chapter will give you a head start in that direction.

SETTING THE STYLE OF YOUR PRACTICE

Life is largely a matter of style. Even though we all face many of the same practicalities and problems, the *manner* in which each one of us goes about handling them reflects our uniqueness.

> Old Professor Jones was drinking tea at a faculty-student gathering. "Here's a lesson in logic for you," he said to Wisenheimer, one of the undergraduates. "If the show starts at eleven, and dinner is at seven, and my daughter has the measles, and my brother plays golf on Sundays only, how old am I?"
>
> "You are eighty-two," replied Wisenheimer promptly.
>
> "Correct," said the professor. "I am amazed at how quickly you arrived at the answer. How did you do it?"
>
> "It's easy," said Wisenheimer. "I have an uncle who is forty-one, and he is only half nuts. You must be eighty-two."

One of the advantages of consulting is that you have the freedom to set your own style and priorities. Direct employees working for large firms do not enjoy this privilege. You can set the style of your practice to be consistent with your personal goals, be they

- maximizing income;
- maximizing free time;
- working only on projects that you find personally interesting;
- working for ecological or humane causes;
- selecting and developing a team based on your personal criteria;

- creating "your" style of work space. The "real you" might want classical music in the background, an office terrarium, office decor that inspires you, or a casual dress code (away from your clients).

DRY SPELLS AND THE VALUE OF AN ONGOING BUSINESS

There is a definite value to having been in business for a number of years. You establish business credibility and a track record of successful projects. If your consulting sees a lull, it is easy to become discouraged. For consultants starting off with only one large contract, the picture has a *binary* character. Things look great when you are "on" and terrible when you are "off."

In a dry spell, remind yourself that the *potential* to land another large contract is always there. You have nearly 100 percent of your time available to *make* that potential come to fruition. But the best preparation for dry spells is to create a *contingency plan* for them. That is, *expect* dry spells and prepare financially for them. Financially, a consultant is like a camel crossing the desert. Without the ability to store water (money), all is lost. Remember, by staying in business, you are maintaining the opportunity to take advantage of future rewarding possibilities.

Dry spells are analogous to the periods between battles for the warrior. They are an exploration of the concept of *waiting*. Some warriors become fat and lazy in the intervals between battles. Others (the successful ones, generally) *use* the waiting period to sharpen their weapons and skills. They prepare themselves by expanding their abilities and strategies to accord with ever-changing developments. The victories are realized on the battlefield, but they originate from the utilization of *waiting*!

When you are in the middle of a dry spell, friends who are not familiar with the nature of consulting can be trying. They will ask you where your next contracts are coming from, and you won't be able to answer them precisely. Your situation is like that of a doctor: You don't know who your patients will be next month; you only know that there is a general level of demand that supports your practice. There are no guarantees. Your knowledge of the business guides you.

Sometimes it's difficult to figure out your next step in a dry spell. If you are able to diagnose the problem behind your lack of business, very well and good. But some problems have subltleties that are hard to understand. Under such circumstances, a quick diagnosis may not be possible. Don't worry, because I have good news: *You don't have to understand a problem perfectly to be able to do something about it!* When you're stuck for a diagnosis, use your resourcefulness, your ability to reach out and ask for help, and your ability to structure experiments. Sometimes, the only way you can determine what to do is to make a few experiments and see which path gives the best results. Understanding arrives after the fact.

Handling dry spells requires a twofold approach. First, develop a marketing action plan. Marketing is a lot of work, yet your time, money, and energy are limited. To become efficient in marketing, you need a plan (as indicated in Chapter 6) that lays out a schedule of phone calls, letters, contact meetings, and promotions to attract the interest of new clients. The emphasis is on action. Don't mope around your office waiting for people to call you; get out there and shake lots of hands! Meet new contacts and prospects.

Getting out there and meeting lots of prospects may not yield instantaneous results, but it surely helps over the long haul. For example, during one dry spell, I found a prospective client who had a pressing problem in my technical area. I spent six hours creating a presentation geared specifically to their need and spent another two hours talking to three of the client's managers about how I could solve their problem. I thought I had made a very convincing presentation, but my efforts appeared to be in vain. Two weeks later, the client told me there was no funding to address that particular problem; they would just live with it. When I heard this, I felt bummed out.

Three months later, I received a call from one of the managers. Was I available to do a very similar project that *was* funded? It turns out that I had already sold myself in terms of abilities. When this new project arose, they remembered my presentation and felt comfortable in giving me the job straightaway. They didn't even ask for bids from competitors; they awarded me the contract on a sole-source basis.

The second part of handling dry spells consists of *emotional inoculation*. That is, learn how to manage your psychological responses to dry spells. Coping with the situation of not having an active contract means:

- Controlling self-deception about how much daily effort you must exert to find new work. Finding new clients and contracts takes eight hours a day. By now you realize that this means marketing multiple clients and writing multiple proposals! You must be more aggressive when the market is down or when your product is approaching the end of its life cycle.
- Overcoming wishful thinking. It's tempting to write just one proposal and wait for a response from that client. However, many clients take a long time to make up their minds. Or they may be off on a two-week business trip while you think that they have already opened your proposal and have asked their purchasing department to write a purchase order. Don't fall into the trap of, "I submitted my proposal; now the ball is in the client's court; it's time to wait." On the contrary: keep on marketing other clients right up to the moment you get a firm commitment in writing.

- Managing disappointment. There are moments in almost every long-term commitment when you may regret your choice. When faced with such regrets, it may help to recall the hassles of direct employment, the unreasonable demands, and the layoffs.

- Countering denial. Acknowledge the true state of the market demand for your current product offering. It's easy to convince yourself that the demand is greater than it actually is. By not denying the truth, you can better face the challenge of creating new or different services to offer.

- Managing worry and depression. These two puppies won't go away by yelling, "Be gone!" But they *will* depart as soon as you start taking action. Worry and depression thrive on the feeling that you have no choices, that you are doomed. Counter these emotions by developing the willingness to experiment with new approaches and see what works. When you're stuck, doing something different—*anything different*—will help you become unstuck. Doing that something different may not be the direct answer, but it will give you the chance to develop a different perspective. From that new perspective, you may see new alternatives or revalue alternatives that you had previously dismissed without much consideration.

DEVELOPING A COMPANY IDENTITY

One potential liability of consulting is that you're always solving *other people's problems.* Year after year of focusing on your clients' emergencies means that you're continually in *reaction mode.* The perfect consultant is always there for the client and centers on the client's needs. This external focus will bring you great success, but it can also cause you to grow stale over a period of years. Eventually, you begin to *think* in reaction-mode terms.

Unless you also start to exercise your own initiatives, you will eventually burn out. The solution is to become *proactive.* Take some time to actively plan your own ventures. Exercise your imagination. Regain the feeling of seizing the initiative. Create new business "experiments."

These comments apply primarily to consultants who have been in business for five years or more. Once you have established a stable client base, it's time to develop products and services that can provide *independent focus* to your business identity.

As an example of how a consulting business can evolve into a product-oriented business, I would like to describe Kaye Instruments Company.

My father started Kaye Instruments Company in 1955 strictly as a consulting company. Drawing on his contacts and reputation as an M.I.T. professor, he was

Tip

Creativity in Business, by Michael Ray and Rochelle Myers (see the Suggested Reading List), offers a wealth of innovative methods to visualize an exciting professional future for yourself. Creative visualization requires the courage to let your imagination roam freely and not be bound by everyday business practicalities. *Creativity in Business* will help spark your imagination and unleash new passion for your business aspirations.

very successful. He hired employees to handle the large research contracts he was awarded. Within a few years, he started searching for a product to make. He and his associates invented a new kind of instrument based on their research on thermoelectricity. They obtained a patent on it and sought ways to produce it commercially.

In 1961, my father died. This left the company of twenty-five employees in a quandary. Without my father's marketing efforts, the amount of consulting they could expect would eventually dwindle. This is exactly what happened over the next few years. During this period, some of the employees worked at consulting while the others toiled diligently to bring the invention into commercial production.

They were successful. Within five years, they dropped the consulting completely and became an instruments company with a worldwide market. Products gave the company focus, continuity, the ability to plan and make projections, a sense of identity, a tangible product, and a profit that was not tied to hourly work rates.

As this book is about consulting, I am not recommending that everyone follow this pattern. My only advice is that you seek a growth path that captures your imagination and prevents you from burning out.

MOTIVATE YOURSELF TO SUCCEED

After extensive research into the art and science of motivation, I have concluded that "pep talks" are a totally ineffective way to motivate yourself. Remember when your boss tried to give you the "rally round the flag, boys" speech to get you to work Saturdays for free? Even though the company was desperately behind schedule and everyone's job was at stake, the pep talk fell on deaf ears. It didn't work because it was phrased *negatively:* "Everybody pitch in and help — or don't bother to come back to work on Monday!"

A much more effective way to motivate yourself is to develop a positive belief system about yourself and your ability to change your life in positive ways. A positive belief system may not be as flashy as the "instant solutions" and "miracle methods" that have been hyped in the motivation industry, but it gives genuine and lasting results.

A positive belief system helps you look for the best in yourself and others. It encourages you to accept yourself as you are. Self-acceptance is especially important for beginning consultants, for it is all too easy to feel intimidated by other consultants who have been working the trade much longer. With faith and belief in your own resourcefulness, you will not be daunted by others who have more impressive credentials or a greater degree of expertise. You'll be able to start consulting wherever you are without feeling inadequate. Once you get going, I'm sure you'll improve your situation.

Here is my ten-step program for motivating yourself in a positive way:

1. **Have a dream.** Having a dream is the sine qua non behind every successful career. It doesn't have to be a grandiose dream, but it must be big enough to fire your passion and induce you to set sail in that direction. In traveling to that far-off destination, do yourself a favor: set yourself steps that are small enough to master in a relatively short time.

2. **Find work you enjoy.** If you don't enjoy what you're doing, having a goal is irrelevant — you'll give up as soon as you hit obstacles. Experiment with different kinds of professional work until you find something worth pursuing.

3. **Run your own race!** Don't compare yourself with others. You don't have to be a genius or a sales whiz or a Nobel prize winner to be a successful technical consultant. Each of us comes with a different mix of talents, credentials, disposition, and resources; it's simply not fair to compare yourself with others in the single dimension of professional stature. Be happy wherever you are. Move to your own music, not someone else's tune.

4. **Look for inspiration in the right places.** Never look to someone who is envious of you or who is competing with you for encouragement. Instead, seek inspiration from those professional friends who want to see you succeed. Also, as part of your annual planning retreat, first read a

motivational book to get your creative juices flowing. See the Suggested Reading List for some good starting points. I find affirmations to be a helpful form of inspiration. Here are some that work for me:

DAILY AFFIRMATIONS

- I enjoy my consulting work and feel confident doing it.
- Being my own boss empowers me to greater productivity.
- I look forward to the challenges and rewards each new day holds for me.
- I am advancing steadily toward my goals.
- My decisions are well considered and reflect my best interests.
- I find it easy to concentrate on the task at hand.
- I am intelligent and persistent in my work.
- I meet new clients with energy and enthusiasm.
- I balance work with family, friends, and recreation.
- I relish the chance to put my marketing ideas into action.
- I believe in myself and my ability to succeed in my work.

---

5. **Celebrate your victories.** No matter how small your victories, make it a point to celebrate them. It means you did a great job! By celebrating, you are acknowledging that fact. When I land a big contract, finish a complicated project, have an article published, or make an especially good business decision, I reward myself with a bonus—some present to myself or a special dinner out with my wife. These rewards are an important motivation; they reaffirm the validity of my goal and my desire to pursue it further.

6. **Don't let anyone knock your dream.** I feel very strongly about this because put-downs are so malicious and destructive. Some people will try to ridicule you by mocking your dream ("A consultant—really! Isn't that a euphemism for someone who's between jobs?") or by holding up a standard for comparison that you cannot possibly meet ("Why aren't you earning $400 an hour like my son the lawyer?"). Naturally, when you start out, you are still experimenting to find the best marketing mix and personal balance. Instead of mockery, you need affirmation that you *can* achieve your dream, that you will indeed discover many ways to better approach it.

Tip

The secret of success in business: It's not how you handle success that counts, but how you handle *failure.* "Winners" suffer as many failures as "losers," but losers run away from their failures and never learn from them. Winners come back from failure determined not to repeat the same mistakes and willing to try things differently until they discover something that works.

7. **Deal positively with failure.** When you stumble, recover as best you can. After the fall, pick yourself up and see what can be salvaged from the incident. Some failures mean that you will lose a customer forever; others can be patched and made as good as new. Whatever the failure, remember to be compassionate to yourself.

8. **Overcome isolation.** Isolation is a major issue for consultants who work from a home office. Unlike a large corporation's offices, your home office can be so quiet and people-free that you begin to develop what psychologists call *stimulus deprivation.* From there, it's only a short step to depression. To avoid this hazard, make sure you get out *every day,* even if only to take a walk or do an errand. Plan to make a few friendly phone calls every day. At least three times a week, meet with friends, contacts, or family to share an activity or a meal. Maintain your personal friendships, for, as a consultant, you will find them even more meaningful and rewarding than before.

9. **Create a support system.** In Chapter 11 I discussed *professional* support in the context of dealing with clients. Here, the support is personal and deals with family and friends. Nurture your family life. Be a friend to your friends. Always remember that friendship and support are maintained by frequent renewal. If you ignore a friend for five years, don't expect him to extend himself very much on your behalf. Always assure that benefits flow to both sides in your friendships, although not necessarily on a tit-for-tat basis.

10. **Take care of your body.** It's really simple. All you need to do is:
    a. Give yourself a proper diet. Don't use food as a psychological means of rewarding or punishing yourself, but establish a balanced diet that agrees with your body. Stay away from junk foods high in sugar or fat—they tend to make it more difficult to concentrate on your work.

HOW TO CREATE A SUPPORT SYSTEM

- Look for support in specific contexts. Don't expect one person to "adopt" you (provide support in all areas of your life).
- Make sure the benefits flow both ways. Reciprocate favors to keep your support system operational.
- Use relationships appropriately. Don't expect someone else to provide you with a sense of identity or assume *your* risks.
- Understand who can support you and who can't. Not all people are interested in supporting you or able to be supportive to you.
- In creating a supportive relationship, pay attention to the unstated part of the deal: Are you comfortable with the other party's ethical standards and unadvertised goals?
- Choose supporters whom you like and personally respect. If you wake up one morning and discover that you're trading support with people you can't stand, it means an evil witch turned you into a politician!

---

b. Exercise every day. Exercise is one of the most neglected motivators you should practice every day. Besides toning your body and making you look and feel more fit, exercise is a wonderful antidote to stress. I know many consultants who use jogging, rowing, walking, basketball, skating, swimming, etc. to forget the stress of the workplace. Take the time to find a form of exercise that agrees with you and set up a regular schedule for it.

c. Get a good night's sleep. Individual needs for sleep vary from four to nine hours per day. Find out how much you need to feel sharp every morning. Nothing looks worse than showing up at a client's office drowsy and disheveled. Clients are quick to make a correlation between "dresses sloppily" and "performs sloppily."

A RADICAL METHOD FOR HANDLING DOUBTS

In my first few years of consulting, I occasionally felt doubts about my viability as a consultant. I recognize now that such doubts are a normal side effect of starting an enterprise.

Learning to live with uncertainty and ambiguity are an important part of self-employment. Unlike the implicit guarantee of "security" that direct employees receive, consultants learn that the only real security is to constantly maintain and improve your own marketability, to run the experiments that show which way leads to the best results. And the way for each consultant is different! Due to the huge differences in our attitudes, experiences, and situations, the paths we tread are wide enough for one person only. That's why trust in yourself is so important — only you can say when it's time to push harder or try a new path.

I have already discussed the importance of positive thinking in handling doubts. Another way to control negative thoughts about consulting is to monitor the alternatives: interview for direct employment! Does this suggestion seem farfetched or radical? *Au contraire!*

If your doubts seem to be getting the best of you, see what direct employment is available, and convince yourself that you are already pursuing your best option. It takes only a small effort to do this, and it will counteract the feeling that you have been closed-minded. It will not detract from your image to quietly and tactfully explore other opportunities. Try not to let your clients know about this, however.

In the event you decide it's in your best interest to rejoin the ranks of the directly employed, I suggest you take out an "insurance policy" on your decision: Do not tell your contacts and clients that you have accepted a direct position until you have been on the job for a month. Do not terminate your health insurance, move your residence, or sell your house for a period of at least two months.

After one month on the new job, you will have gained enough of an impression to judge whether the promises made by your employer were trustworthy. If your employer misrepresented important aspects of your new position, you may have no real leverage to remedy the situation. In most cases, you will still be a hired hand, even if you have an impressive title, salary, and office. You can discuss differences with the management, but you must remember that it is *their* company. Even if you have an employment contract defining the scope of your authorities, responsibilities, and mutual obligations, in practice it is extremely difficult to address a company *legally* for stretching the truth.

After your first month on the job, consider whether you made a good decision. The "insurance policy" makes your decision to leave consulting *reversible.* If you have been the victim of an unethical practice (or have

MEET THE PROS UP CLOSE AND PERSONAL

Dan Popok is a mechanical engineer who specializes in heat transfer analysis. After receiving an M.S.M.E. from North Carolina State University in 1983, he went to work for a large industrial firm. In 1994, soon after he earned his P.E. (Professional Engineer) license, his employer started having business prob-lems. Dan says, "I was working in a place that abounded with technical talent but that was lacking in good management. We were reorganized frequently. As the company downturn progressed, we were pressured more and more to write proposals on our own time. This resulted in hastily assembled proposals that weren't good enough to win contracts. Management started preaching about the need to do more work uncompensated, but never took corrective action to prevent the need for uncompensated work in the first place."

Dan evolved his own "density theory" to explain how authority floats upward and responsibility sinks downward in most vertically managed organizations. The density of authority is lower than that of responsibility, and the two are immiscible. Inevitably, management winds up calling the shots but delegates the responsibility for getting it done downward. If the call is bad, management blames the responsible subordinate, regardless of whether the subordinate was given sufficient authority to get it done or if it was a bad call in the first place.

In this situation, Dan felt "adrift, with little chance to alter the course of things. With no new projects on the horizon, there was little incentive to try hard. Morale suffered as survival inevitably meant stretching out tasks. After all, if you're already paddling your canoe toward a waterfall, how enthusiastic can you get about being asked to row harder toward the drop-off?"

Dan decided to try some moonlighting consulting. Due to lack of work, Dan's employer started permitting half-time work, just as Dan's consulting opportunities began to appear. Dan's practice went so well, that he now consults full time. He says, "I feel a lot better about myself as a consultant. I have clients of my own choosing, and I'm the one who reaps the rewards of my hard work. In independent consulting, there are no butts to cover, except my own. And I'm fully empowered to cover it myself, or to keep myself from getting into scrapes where it needs covering in the first place. I'm at the helm of my ship, and if I crash it into the rocks, it's my own doing. Even if the seas are choppy, I feel better with me at the helm than with someone I have no faith in."

<u>Warning</u>

You're fooling yourself if you think you can outsmart an employer by asking for an employment contract that "guarantees" your advantages. Being someone else's employee means that the employer is the one in control. The employer has more options, moves, and countermoves than you could possibly anticipate in an employment contract. The only real protection against an employer's exaggerated claims is to keep *your* options open (i.e., structure your moves to be as reversible as possible) until it is clear that you made a good decision.

merely been shortsighted), you can safely recover your consulting practice without loss of momentum.

Is this method of handling doubts ethical? Am I making a virtue of mistrust? It would appear that once a decision has been made, either to enter consulting or to leave it, it should be carried out in good faith. But the problem is that we change and situations change. A blind consistency benefits no one, whereas maintaining one's options is always wise and safe.

Therefore, what I am really recommending is that you monitor your options and make sure you don't wander too far from those that are best for you. In my years as a consultant, I have "given up the ship" twice for direct employment. The positions looked very attractive from the outside, that is, before I started the job. In both brief stints, I found that I could not have anticipated the negative factors that made the positions undesirable. However, in both situations, major shortcomings were obvious after just one month.

I know many acquaintances who have been duped or misled in employment interviews. It is an industry-wide problem. Do not let someone else's exaggerated claims threaten your happiness and prosperity.

After a few years of consulting, you will probably find the impulse to consider direct employment weakening. Your track record of consulting successes will speak louder than the promises of employment recruiters. Results are hard to ignore!

AN INVITATION TO WINNING

Now we come to the fun part! Visualize yourself winning that game of tennis! And visualize yourself *winning* in consulting: Every new project is an extension of your interests. Clients want you to solve problems that interest you as well as them. Each new assignment furthers your career.

When you reach the point where the customer wants *you,* and no one else will quite do, you have arrived. You will find yourself being paid top rate, rather than being sent down to the purchasing agent for "trimming." ("Gee, all the other consultants we use are charging only $25 per hour. Could you meet this figure?") At the end of the month, when the client checks start arriving in the mail, you are amazed. "Wow! Look at all that money! This sure beats the rat race of direct employment."

You are able to take enough time off to plan your moves thoughtfully, and you pick those that benefit you most. Just as the experienced tennis player knows how to judge the amount of spin on his opponent's serves, you know the game well enough to perceive a losing project and learn to decline graciously. You develop your own "bag" of serves, which means that you know how to generate interest in new work that gives you a unique edge over your competition.

You start to use your creativity and imagination to enhance your work. You gain the confidence that comes only from the experience of transforming a dream into a tangible reality. You develop niches and products that are your trademarks. Work isn't work anymore; it's play for money!

EPILOGUE

WHEN HAVE YOU "MADE IT" IN CONSULTING?

How do you know when you've "made it" in consulting? Is it after you have survived for one year or three? Is it after you have made your first million dollars or hired your hundredth employee? Is "success" as distant as winning the Nobel prize or being admitted to the National Academy of Engineers?

It seems strange, but my definition of "making it" has changed and evolved as the years have passed. When I started out, *survival* was success. After three years, "success" was to earn as much money as I could. And now "success" is using the resources at my disposal to "play my game," to lead into my professional interests rather than away from them. The definitions of success go hand in hand with the hierarchy of goals described in Chapter 15.

It is easy to underestimate the amount of effort needed to organize and establish a consulting business, or almost any business, for that matter. Many people tell me that they would like to be in business for themselves, but they can't take the time or accept the risk. Well, like most things in life, nothing is won by sitting back and daydreaming.

ENCOURAGE YOURSELF BY ASSOCIATING WITH OTHER SUCCESSFUL PROFESSIONALS

The greatest obstacle to starting my own consulting practice was figuring out that I could really do it. A hundred doubts haunted me as I ventured into my own business. I knew I could handle the technical part of the enterprise, but could I also be a good businessperson? I had little support, as most of my friends were direct employees who had a "factory mentality." They could not conceive of anyone leaving the "security" of corporate life. Professional independence was not a reality for them.

I needed more contact with other consultants and businesspeople to learn that it *could* be done. The importance of seeking out a circle of consultants and self-employed professionals for socialization and motivation cannot be overemphasized. The psychological pressures of consulting are such that it is essential to have a group of people who share your situation and support you. When you have suffered a professional loss and need someone to talk to, it helps enormously to speak with a person who has gone through the same ordeal. He or she is much more likely to offer genuine compassion. He or she often suggests alternatives that you never considered. I have gained much over the years from fellow consultants who have said, "Why, here's what I did when I was in that dilemma."

WHAT I'VE LEARNED IN CONSULTING

When I started consulting, I thought the primary resources I would need were a good technical library and lots of money. I now understand that many less tangible resources are required, such as the desire to work hard and make a dream come true. These resources and energies are every bit as important as items that show up on inventory and balance sheets.

In my first two years, I had no idea that marketing and management activities would be so important. I focused on my technical abilities, erroneously thinking that they would carry me through all my problems. I had carried over a premise that was sometimes true in direct employment, but always false in consulting. Technical ability is not the whole story!

After I developed a feeling for the larger context of engineering work, with its business, management, and social implications, I was not satisfied with seeing everything in terms of my narrow specialty. My desire to specialize in graduate school was partially motivated by the prejudice that business was less pure and less desirable than the ivory tower of science. This

was a common attitude among engineering students —a disdain for business based on the incorrect view that it is the greedy pursuit to outwit others in mercenary deals.

As I gained experience, my view of business changed. Business identity is what *allows* a person to determine the way he does his work: the projects selected, pace, style, and rate of pay. At the beginning of my career, I had naively thought that these "trivialities" would be taken care of by my employers. These concerns turned out to be nontrivial, and my employers had no interest in attending to them on my behalf!

There is no "lazy man's way" to take responsibility for your own business identity!

Many of the qualities that are helpful to the successful consultant are also helpful in making a person a successful direct employee. I hope this book has been useful in contrasting the two situations and in presenting consulting as a realistic alternative to direct employment. If you are suffering as a directly employed engineer, perhaps you are fighting the wrong battle. Consulting may be the thing for you.

On the other hand, many engineers and technical specialists are actually better off as direct employees. Although they may be every bit as talented and motivated as those who venture into consulting, the rewards and challenges of self-employment may not be their cup of tea. The choice is yours.

In writing the first and second editions of this book, I emphasized that prospective consultants *should* satisfy certain requirements before venturing into consulting. I do not wish to downplay the value of Table 18 in Chapter 17 as a decision aid. Nevertheless, over the years, many letters from readers have convinced me that attitude, determination, and a good strategy are often more important than scoring high on the Consulting Qualifications Quiz.

No matter what or who you are—young/old, technical expert/novice, marketing whiz/beginner—I hope this book has shown you how to *build* on what you already have to achieve more success. Whatever your present limitations, you can make up for them with a positive attitude and by creating strategies to overcome the obstacles you face. If you lack start-up money, don't worry—you may be able to bypass that limitation by following the advice in Chapter 12. If you lack credentials, don't fret—you can build them up with the strategies presented in Chapter 4. If you lack extensive marketing experience, don't be put off. Instead of sitting on the sidelines, review Chapter 5 and put some of its marketing ideas into motion. If you have been laid off or given an early retirement, don't despair—Chapter 6 shows how to repackage yourself as a consultant and highlight the very best you have to offer. And please, when you attain the success you so richly deserve, write me and let me know how you did it.

NO ONE CAN DO IT FOR YOU

There is an old Zen story about the monk Joshu. After twenty years of serving his master, he had finally become enlightened. The master was very old and approaching death, and wanted to test his trusty disciple. Inviting Joshu into his room, he explained that he would be dying soon. The master said, "You are the most advanced of my disciples. After I go, you shall be the master. You are ready."

The master handed Joshu a thick book, saying, "This is the book my master gave me forty years ago. It has all the secrets of Zen meditation and enlightenment. Read it carefully and it will serve you as well as it has served me."

Joshu took the book from the master and immediately threw it into the flames of the open fireplace.

"What are you doing?" cried the master.

Joshu replied, "The only help I need comes from within."

The master exclaimed, "You are indeed worthy to become the master, for there was nothing written in that book. It was my final test to see if you would fall for a lazy man's way, if you would value someone else's effort more highly than your own. Now I can die with a peaceful heart. I know you will carry our tradition."

The point of this story is that no one can do it for you. Whether it is Zen or consulting, the efforts have to be your own.

THIS IS NOT A GET-RICH BOOK—IT IS A GET-PROFESSIONAL BOOK

When I was writing the first edition of this book, I stopped after completing 90 percent of it. I was afraid that it would be confused with the plethora of "get-rich" books on the market. It remained unfinished for two years. One day I finally realized that this is not a "get-rich" book, but a "get-professional" book. It is presented to the reader in an attitude of genuine service. I truly hope that you will be able to benefit from my experience in starting your own consulting practice.

Instead of "good luck," I wish you "good effort"!

Appendix: Case Study

This case study is a chronological history of a recent consulting project. It illustrates many of the issues discussed in Chapters 5 through 12. All company names and details have been changed to protect the privacy of the client.

FIRST CONTACT

For two years, I was a visiting professor at the University of Massachusetts, where I taught fluid mechanics and thermodynamics to juniors and seniors. One day early in February, I was in the department mail room, which was the informal faculty gathering place. The copying machine was located there, and I was making overheads to use in my thermodynamics class. Peter, who taught the design classes, stopped to chat while I was working. I did not know Peter very well; we had talked only briefly before. After exchanging pleasantries about the New England weather, Peter said, "Somebody told me you know a little about fluid mechanics."

I kidded back, "Well, you know how it is. A little here, a little there. This is my third term teaching fluids, and now I almost understand it."

Peter winked at me and continued, "Seriously, though. I've got a client who needs some help with a fluid mechanics problem, and people here say you're the one to ask. Are you available?"

"Sounds interesting," I said. "Tell me more."

"Well, my client makes instruments that use fluid manifolds, and they're having difficulty getting one of them to work. I don't know much more than this. Can I give them your card?"

"Of course," I answered, pulling a business card from my wallet. "How can I get in touch with them?"

Peter replied, "When I speak to them next week, I'll give them your card and have them call you to discuss the problem. They've been a good client of mine, and I'm sure you'll enjoy working with them."

The next week, I received a call from Matt, the project engineer at Universal. He wanted to talk about the fluid manifold problem. I quickly grabbed pencil and paper. (Taking careful notes on client inquiries is immensely useful. I never fail to do this.)

Matt described the device his company was designing and briefly explained its operation. As Matt talked, he mentioned flow rates, gas compositions, and construction details that meant absolutely nothing to me. Since Matt had been immersed in this project for a long time, he "forgot" that I had never seen his machine, which was unique. Therefore, I tactfully asked him to explain the few items I needed to grasp the nature of his problem. The lights went on. I could see that he needed someone to analyze the device and recommend a fix that would allow it to operate properly.

I mentioned that I was very familiar with that kind of fluid distribution analysis and was sure I could come up with a solution. It might, I offered, involve a short computer program to calculate the flows, which I could also write for them.

Matt wanted to know my experience in designing manifolds and whether I had worked on gas manifolds in particular. He was under the impression that the design of gas manifolds was inherently different from the design of liquid manifolds. I explained that for low velocities, as existed in the present situation, the flow was incompressible, even though it involved a gas. I described the technical reasons for this fact. (That was a piece of cake, since it drew on lecture number three of my Fluids 101 course!) By the time I finished my explanation, Matt was convinced I knew enough to handle the problem. I concluded by saying that I had indeed designed many manifolds for gas flow.

(Matt, an electrical engineer by training, is typical of many clients. The information explosion has proceeded to the point where technical people no longer have a large common knowledge base. It is a mistake, therefore, to assume that clients will know even the rudiments of your technical specialty. I don't look down on my clients for their lack of specialized knowledge. The shortcomings are reciprocal: I am grateful Matt didn't ask me about his electrical engineering!)

Matt then inquired about my charge rate. When I answered him, he said, "OK. How long will it take to get results?"

I told Matt I wasn't sure at the moment, but that I would think about it for a day and write a short proposal. (By declining to answer immediately, I avoided a hasty estimate that I might regret later.) Matt was receptive to the idea and gave me his address and phone number. He mentioned that he would like to get moving on the project, so I promised to send the proposal within two days.

TABLE 20. Project Schedule

Task	Hours	March	April	May
Kickoff meeting	5	X		
Review drawings	6	XX		
Construct flow model	10	XX		
Calculate model properties	10		XX	
Create computer program	10		XX	
Verify computer program	4		X	
Determine input	6		XX	
Interim report	2		X	
Run computer model	8		X	X
Do parameter studies	8			XX
Meeting to decide fix	5			X
Final report	6			XX
Total hours =	80			

Finally, I asked Matt about the project's time frame. He told me he needed answers in two or three months at the latest. "No problem," I responded. (This meant I would be a busy puppy for the next two months, considering my teaching schedule and other commitments.)

The next day, I sent out the proposal. It was a one-page "letter proposal" commonly used for small jobs. It had a five-line statement of work, a short section on cost and delivery, and billing terms. I also enclosed my consulting brochure and three portfolio sheets showing manifolds that I had designed.

To make the cost estimate, I prepared, but did not send out, a Gantt chart. (See the project schedule in Table 20.) By comparing this chart with my existing work schedule, I assured myself that I could handle the work load.

The ball was in the client's court. I waited two weeks. Nothing. I called. Matt said that my consulting contract was in the approval stage. I should receive a call from purchasing in a few days. When that came through, he would call and set up the kickoff meeting.

The next week, a woman from purchasing called with a verbal go-ahead including a purchase order number and verbal confirmation of my rate and the total dollar amount. She said the paper version of the purchase order would reach me in a week. (I thought, "This is easier than most of my projects. Marketingwise, Peter paved the way for me. I have not had to do any selling. His intervention enabled me to bypass the stages of establishing contact and credibility.")

Matt called the next day. We set up a kickoff meeting date, and I asked for directions to the Universal plant, as it was a hundred miles away.

At the kickoff meeting, Matt brought me to the conference room. On the way, he introduced me to the "big boss" who approved consultant contracts, but who didn't get involved in technical details. At the meeting, there were three other Universal engineers. We introduced ourselves, shook hands, exchanged business cards, and poured ourselves some coffee.

We got down to business. Matt showed a prototype of the unit and explained what he thought the problem was. The other engineers added more data and information about the unit's operation. I took copious notes but tried not to overdo it. (At a first meeting, pay some attention to the people dynamics to make sure you are maintaining rapport.)

During the next two hours, I asked dozens of questions and gathered enough data to start the project. At the end of the meeting, I outlined my approach to the problem. It met with everyone's satisfaction. I left with the prototype and a stack of engineering drawings four inches thick. This was a very complicated device!

HOW TO UNDERSTAND THIS DAMN THING?

I started work on the project, following my schedule. The first task was to go through the pile of drawings to understand how the manifold was built. It was discouraging at first, because there was *too much* information. I needed to decide which drawings were necessary for my purposes. This took considerable effort. Fortunately, I found out that disassembling the prototype aided immensely in visualizing the complicated three-dimensional geometry.

The next week, I called Matt to clarify an inconsistency between two of the drawings. (I would have called anyway, even if I had had nothing substantive to discuss. It is good policy to maintain frequent contact at the beginning of a project. If there is no pressing reason to call, "invent" a few questions to serve as the excuse.)

By now I was developing a better feel for the device and how it operated. I made some preliminary calculations to make sure that I could "ballpark" the manifold's flow characteristics. Once satisfied, I moved on to modeling each small component in the flow path. This took ten hours of detailed calculations. I pushed forward, realizing that it was a necessary step in creating the computer model.

DISCOVERY TIME

For the next week, work continued as planned. Then I made a discovery. Everybody—including myself—had been calling this unit a *manifold*. After studying it, I concluded that it was not a manifold, but a *fluid network*. It's more than just a difference in words; these two animals behave differently. I asked myself whether this discovery would affect my plan of action. Yes, it meant writing a computer program for a network instead of a manifold. This is more complicated. I hoped I could absorb the extra work without having to ask for more money.

I called Matt to tell him about my discovery and to inform him of my progress. To my surprise, Matt was not very concerned. He asked, "Can you handle it?"

"Yes, no problem," I assured him. (Over the next few days, I redoubled my efforts to make sure that *I* believed there was "no problem.")

At the end of the month, I sent out my first invoice for work completed to date.

Two weeks later, my efforts started to pay off. I had firmed up enough material to issue an interim report. It contained a nice picture of the flow model done in CorelDRAW! (a computer graphics program), an Excel spreadsheet showing flow properties for all the components, and a picture of the network model. The cover letter stated that I was diligently writing the network program in QuickBASIC.

When Matt received my interim report, he called to say he was pleased with the progress and to ask when I would have results.

PLOTTING A SOLUTION STRATEGY

By now I had enough insight to see exactly how to approach the problem technically. I made a detailed list of the remaining steps.

- Construct six different spreadsheets to represent varying flows as a parameter.
- Figure out how to write the file input/output routines needed for the computer program.
- Create five verification cases from the literature to make sure my program gives the correct answers.
- Make a simplified network model to understand the gross behavior of the manifold.

- Make the detailed model to arrive at the final answer.
- Compare my results with the test data Universal compiled.

One by one, I polished off these tasks. It was more work than I thought, but when Universal's check for the first invoice arrived in the mail, my spirits were renewed.

WHO SAID IT WOULD BE *EASY*?

As I continued with the project, I was not pleased with the numerical values of my results. Since I had already verified the program, I concluded that the program *inputs* must be incorrect. (As the cliché about computers goes, "Garbage in, garbage out.") I must have made a mistake in calculating the loss coefficients of some components. I checked my formulas and math again but could not find an error.

I decided to examine the losses at each point in the flow path and see if they made sense. This strategy paid off, for I found one spot where the velocity was extremely low. Although very few books on fluid mechanics mention this fact, when the velocity becomes that low, a different loss coefficient is required. I tried plugging in different values for this coefficient, and it had a significant impact on the results.

Then the only problem was that I had no experimental or theoretical data for this low-velocity coefficient. It was for a very unusual geometrical configuration. I searched my own library without luck. The university library did not contain this data, either. I located only an indirect reference to an obscure reference book. Neither I nor the university owned it. Defeated. Then I remembered a colleague who might own the book. I called Bill, and sure enough, he had the book. Bill owed me a favor. I convinced him to loan it to me for a week. When I finally inserted the correct coefficient, all my calculations fell into place. Success!

For me, it is a challenge to assess a technical problem at its proper depth and come up with the solution. Research has always been fun for me, and I enjoy the chance to dig a little deeper. However, I must remind myself not to spend too much time refining my answers, or I will overrun the allotted time. I was concerned about the extra hours I had spent in tracking down the oddball loss coefficient. The next time I spoke with Matt, I mentioned the possibility of adding eight hours to the contract if I ran over. He said OK, but added that I should wait and see if I really needed it.

Technical ability and experience are important factors in consulting success. That nine out of ten textbooks cite a loss coefficient without stating the limits of its validity is not sufficient excuse to botch a project. Regardless of your technical specialty, recognize when you need more data and figure out how to get them. Your innate resourcefulness will help you track down the literature, equipment, or suppliers you need to skillfully complete the project.

NEGOTIATING A USEFUL ANSWER

Once the computer model was working, it was easy to figure out which parts of the device could be changed to achieve the desired performance. I performed an analysis to see which changes ("fixes") were the simplest and most effective. With this data in hand, I called Matt and got his offhand reactions to the three most promising alternatives. Matt was very receptive to one option that involved only a minor modification to the existing design. I also felt that this alternative was their best bet. However, instead of deciding right then, I suggested a group decision-making meeting at Universal. I promised to bring enough data cases so that we could consider the most promising options.

To ensure that your recommendations will be accepted, get the *group* to make the decision. If you are sitting on top of the data, there is little to lose with this approach and much to gain. First, you will be more familiar with the analysis and trade-offs than anyone else. It is unlikely that one of the client's staff will strong-arm you into an unfavorable technical position. Second, asking for group input prevents "showstoppers" from emerging later (managers of groups such as safety, quality, manufacturing, and purchasing finding reasons to shoot down your recommendation). Third, this is your best form of advertising. It is your time to shine. In the group decision-making process, you appear as the "guru," the knowledgeable one. Play it right, and you'll gain favorable exposure to the client's management.

As a policy, I usually call to confirm meetings two days in advance, especially if I must travel to get there. It's a good thing, too, because Matt said he couldn't make it. He forgot to call me to set up a new meeting. We agreed to postpone to the next week. As we said good-bye, Matt expressed his satisfaction with the way things were going.

At the end of the month, I sent out the second invoice.

When we met at Universal, I presented lots of data. I had prepared summary overheads to show the current performance of the unit and the performance for the alternative fixes. Supplementary material was available to describe the side effects of each fix and the cost and manufacturability implications.

The meeting lasted two hours. At the end, everyone agreed with the fix that Matt and I recommended. Everybody was satisfied. Matt instructed one of his engineers to start implementing the fix. Matt smiled and thanked me. As I left, he asked about the final report. I told him I'd get right on it.

PREPARING THE REPORT

It was time to write the report. This was easy, because I had already completed the calculations and the client had agreed to the proposed fix. Why write the report? Well, until now, the client had only *heard* about my calculations. They had presumed—on good faith—that I had made my recommendations on a technically defensible basis. When they received the report, they would see my calculations and *know* that this was the case.

The report started with a one-page executive summary that briefly stated the problem, described my approach, and gave my results. I concluded by noting that the fix solved the client's flow distribution problem. On subsequent

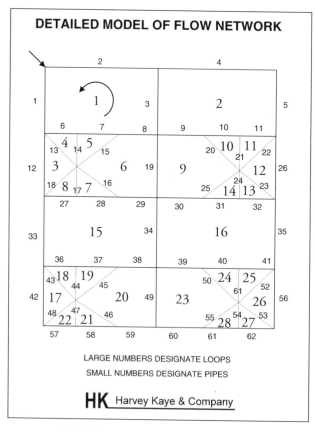

FIGURE 9. Illustration of flow network.

pages, I gave supporting calculations, details, and three computer graphics showing the flow geometry. (See Figure 9 for an example.) In an appendix, I placed printouts of the twenty-five computer runs spanning the range of parameters. I also enclosed a listing of the QuickBASIC program. All in all, the report was thirty-three laser-printed pages.

FOLLOW-UP

A week after I sent the final report, I called Matt to see how things were going. He was pleased with the report and my involvement in the project. Matt told me to mail my final invoice.

Three weeks later, Matt called and asked me to attend a meeting the following week. His engineers would be discussing how the changes I had recommended would affect the fluid sealing design. Matt wanted me to be present at the meeting and to offer in-depth support for the fluid mechanics aspects. Matt felt this was a one-time deal and said to consider this an extension of the purchase order that we had just completed. Rather than write a new order, he would have purchasing add half a day's charges to the Not-To-Exceed amount. Since I had already received the check from my second invoice, I felt comfortable in accepting his verbal go-ahead.

(From my viewpoint, the financial risk in complying with his wish to forgo official paperwork was small. Further, it was wonderful to be paid to attend a meeting where one possible outcome was another contract.)

At the meeting, I met two engineers from another group at Universal. I gave them my card, and we sat down to discuss the problem. After a half hour of asking questions, I was able to cast the technical issue into a classical model. A simple back-of-the-envelope calculation took ten minutes and resolved the problem on the spot. Everyone was pleased. I drove home confident that they would call me again when they needed fluid mechanics help.

Four weeks later, Peter and I were sitting in a restaurant at the university. I asked him out to lunch as a way of thanking him for the contact and to get to know him better. After the usual jokes and complaints about university politics, Peter loosened up.

"Harvey, I've had my fingers crossed these past few months. Originally, Matt gave your fluids problem to one of the senior mechanical engineers in his group. The engineer spent two months fumbling the analysis and was not even close to an answer. Matt finally took him off the job and decided to hire a consultant. I'm very relieved you were able to solve the problem."

"Gee, Peter," I replied, "if I had known that, I would have estimated the job at three weeks instead of two. C'est la vie."

"Don't worry about it. They like what you've done so much, I'm sure there will be other chances. You know, I've done consulting part-time for twenty years myself, and I love it. The money is great. It's always challenging. And you meet lots of interesting people. At times I wonder how *anyone* could be happy as an employee when there are such great opportunities in consulting ...This is the good life."

I held up my glass to his and said, "Hear, hear!"

Suggested Reading List

Thomas H. Athey, *Systematic Systems Approach.* Englewood Cliffs, NJ: Prentice Hall, 1982.

This book on problem solving is the most readable and interesting I have come across. It clarifies the relationship between problem formulation, establishment of alternatives, information requirements, constraints, criteria for solutions, and evaluation processes. The book states the principles clearly and illustrates them with colorful examples. This book is valuable for consultants who work with systems that are sensitive to financial, political, environmental, and legal conditions.

Edward de Bono, Tactics: *The Art and Science of Success.* Boston: Little, Brown and Co., 1984.

Edward de Bono has achieved worldwide renown as a creative-thinking expert. He originated the concept of *lateral thinking,* in which creative solutions are found not by digging deeper in the same hole (logical, linear mode of thinking), but by digging different holes (metaphorical, nonlinear mode of thinking). In this book, he examines the tactics and strategies used by many successful people throughout history. I like his distinction between general and detailed strategies. Of the former, de Bono mentions two types: either you structure yourself against failure by erecting barriers to limit your losses, or you structure yourself to maximize success.

Some of the topics covered by de Bono are opportunity seeking, opportunity creating, generating new ideas versus a "me too" approach, whether you must take risks to succeed, creating strategies, how rigid a strategy should be, tactics, knowing your opponent, the merit of surprise, the proper place of tactics, and the significance of intuitive thought.

Julie K. Brooks and Barry A. Stevens, *How to Write a Successful Business Plan.* New York: AMACOM, 1987.

Although many books on business planning are available, most are geared toward manufacturing and retail operations. This book is general enough to be useful to consultants. Brooks and Stevens discuss defining your business goals, defining your "product," analyzing the market, positioning yourself within the market, determining your capital needs, sales forecasting, financial analysis, creating milestones to measure your progress, and more. They emphasize that the type of business plan you should write depends largely on your intended audience, be it a bank, a venture capital company, a group of partners, or yourself.

353

William T. Brooks, *Niche Selling: How to Find Your Customer in a Crowded Market.* Homewood, IL: Business One Irwin, 1992.

> This book helps you find your way in the crowded markets of corporate America. According to Brooks, *focus, leverage,* and *alignment* are the conceptual tools that enable you to compete effectively in a crowded and rapidly changing market. By focusing on your customers' needs, you can find new niches in which you have a competitive advantage. By leveraging your efforts, you can discover ways to adapt your "product" to new trends and introduce them to the market before anyone else. By aligning yourself with your customers and their evolving needs, you will be able to give better, faster, and more effective service. *Niche Selling* covers the middle ground between sales and marketing. This dynamic book offers a wealth of tips and pointers to anyone who wants to master the art of selling. Highly recommended.

Theodore A. Rees Cheney, *Getting the Words Right: How to Revise, Edit and Rewrite.* Cincinnati: Writer's Digest Books, 1983.

> Cheney's guide to becoming a more skillful writer uses a three-step editing "system": revising by reduction, revising by rearranging, and revising by rewording. The section on rearranging shows how to rethink the logical order of your material. It helps you ensure that relationships are clear, that transitions are smooth, and that your report has coherence from beginning to end. Cheney shows how to emphasize your major points and how to overcome problems of shifting point of view. If you apply his techniques, your writing will be leaner, more logical, and more persuasive. *Getting the Words Right* reads like a novel; it is a thoroughly enjoyable way to improve your writing style.

John D. Constance, *How to Become a Professional Engineer,* Fourth Ed. New York: McGraw-Hill, 1988.

> This book offers valuable information about getting your Professional Engineer (P.E.) license. Constance is well qualified in this subject, being a P.E. himself and having written extensively on P.E. registration and review methods for P.E. exams.

Sarah and Paul Edwards, *Secrets of Self-Employment: Surviving and Thriving on the Ups and Downs of Being your Own Boss.* New York: G. P. Putnam's Sons, 1996.

> This book describes the psychological aspects of self-employment and offers useful advice for coping with the many situations you will encounter. Although it's addressed to self-employed individuals in general, much applies to self-employed consultants. I especially liked the chapters on motivation, enduring the "emotional roller-coaster," and making the mental shift from direct employment to business independence.

Ernst & Young, LLP, *The Ernst & Young Tax Guide* (Issued Annually), New York: John Wiley & Sons.

> This tax guide explains the federal tax code in language I can understand. With clear, concise exposition, it takes you step-by-step through all the forms you'll need to fill out for your individual return. It also has a useful chapter on Schedule C returns for the self-employed. *The Ernst & Young Tax Guide* is chock-full of little sidebars called *TaxAlerts* and *TaxSavers* that clarify important issues and show how far interpretation of the IRS rules can be stretched. The clear examples are very helpful.

William C. Giegold, *Practical Management Skills for Engineers and Scientists.* Belmont, CA: Lifetime Learning Publications (Wadsworth), 1982. Reprinted by Krieger, 1992.

> When you are a consultant, you are your own manager, like it or not! Giegold's book is a superb treatment of the four basic management functions: planning, organizing, integrating, and measuring. This book is useful not only for your own self-management; it gives insight into how your clients' management may (or may not) be working. The author's enthusiasm for his subject shows on every page. Giegold deals with the *essential,* as opposed to the *incidental,* aspects of management. I found his book full of applicable advice and insight.

Victor G. Hajek, *Management of Engineering Projects,* Third Ed. New York: McGraw-Hill, 1984.

> If your consulting involves project management, this book provides valuable background information. It succinctly treats proposals, estimates, contracts, project monitoring and scheduling, and project performance criteria.

James M. Higgins, *101 Creative Problem Solving Techniques: The Handbook of New Ideas for Business.* Winter Park, FL: New Management Publishing Company, 1994.

> This short and lively book offers an abundance of fresh ideas on problem solving in a business context. The author places considerable emphasis on coming up with new and useful alternatives. This is the battlefield on which most problem-solving campaigns are won or lost. The better the set of alternatives to choose from, the better the solution will be. Toward this end, Higgins offers thirty-eight techniques for individuals to generate more alternatives, and thirty-two techniques geared to group generation of alternatives. Highly recommended.

Ron Hoff, *"I Can See You Naked."* Kansas City, MO: Andrews and McMeel, 1992.

> This national best-seller is a witty and irreverent guide to making memorable presentations. In a colorful and anecdotal style, Hoff shows how to prepare your presentation to suit the audience, how to deliver it with flair and panache, and how to make sure that people don't fall asleep while you are talking. If your consulting involves giving frequent presentations, the tips, insights, and techniques in this book are sure to help.

Herman Holtz, *Expanding Your Consulting Practice with Seminars.* New York: John Wiley & Sons, 1987.

> Holtz's book is an excellent resource for anyone who wants to offer seminars as a complement to consulting. Holtz discusses the basic kinds of seminars, how to plan and advertise, how to set registration fees and make income projections, materials and equipment, arranging times and location, and, very important, how to convert seminar attendees into regular clients.

Tom Hopkins, *How to Master the Art of Selling,* Second Ed. New York: Warner Books, 1982.

> Do you sell the steak or the steak's "sizzle"? To find out, read this best-selling book. Tom Hopkins, a master sales trainer, offers a useful introduction to prospecting, sales

psychology, dealing with customers, getting the first sales meeting, arousing customer interest, and closing the sale. He gives comprehensive advice for many different selling situations and peppers the discussion with examples and colorful anecdotes.

Internal Revenue Service, *Tax Guide for Small Business.*

Issued yearly as IRS Publication 334, available free from your local IRS Forms Distribution Center. This guide is "must" reading, as it discusses the tax aspects of operating a small business, including record-keeping requirements, acceptable accounting methods, allowable expenses, and tax returns. It is also available in electronic format from the IRS Internet site (http://www.irs.ustreas.gov).

Harvey Kaye, *Decision Power: How to Make Successful Decisions with Confidence.* Englewood Cliffs, NJ: Prentice Hall, 1992.

How can you tell which choices will bring you the greatest personal satisfaction? Read *Decision Power.* It will guide you through the issues and help you identify the risks, rewards, and trade-offs involved in becoming a technical consultant. *Decision Power* shows how to combine the wisdom of your rational mind with the strength of your emotions. It helps you overcome obstacles and encourages you to say yes to the decisions that meet your real needs. This book shows you how to make difficult decisions without letting fear of failure or fantasies of grandeur lead you astray.

Genie Z. Laborde, *Influencing with Integrity.* Palo Alto, CA: Syntony Publishing, 1988 (Available from the publisher at 1450 Byron Street, Palo Alto, CA 94301).

Your success in sales meetings and person-to-person interactions is largely dependent on your communication skills. Part of the communication process is getting feedback from and gaining rapport with the other person. *Influencing with Integrity* is a modern psychological approach to developing these valuable skills. For example, what clients say to you is often not what they are thinking. Laborde's hints on reading nonverbal messages are extremely helpful in deciphering the "real" message.

In this book, Laborde exemplifies the very communication skills she sets out to teach. With pictures, analogies, anecdotes, and simple exercises, she "reaches" the reader. Her method is based on the premise that the communication is not achieved until you get the response you desire from the listener. If the response is not what you want, it is up to *you* to change the message (or the manner of the message) until you get the desired response. This differs from the common approach to communication, which is to repeat the same message louder and think, "My message is perfectly clear; what's the matter with *you*?"

John M. Lannon, *Technical Writing,* Fifth Ed. New York: Harper College, 1991.

This excellent six-hundred-page text covers all you need to know about writing reports, proposals, and business letters. It teaches you how to organize, outline, summarize, and revise for maximum clarity and impact. Lannon gives methods for improving descriptive writing, technical explanations, and research. He also offers advice on word choice, tone, and sentence structure.

Jeffrey Lant, *The Unabashed Self-Promoter's Guide,* Second Ed. Cambridge, MA: JLA Publications, 1992.

> The subtitle of this book says it all: *What Every Man, Woman, Child, and Organization in America Needs to Know About Getting Ahead by Exploiting the Media.* Lant's punchy guide to public relations shows you how to use the public media (newspapers, journals, radio, television, etc.) to gain greater professional exposure. Some of his techniques will be useful to consultants.

Edwin T. Layton, Jr., *The Revolt of the Engineers.* Cleveland: The Press of Case Western Reserve University, 1971.

> Reprinted by Johns Hopkins, 1986. This is a stimulating discussion of the incompatibilities between the *professional* and *business* aspects of engineering. Layton explains that *professionalism* is more than identification with a specialized body of knowledge. Professionalism also carries the ideal of autonomy, the privilege to set one's own pace and apply one's own judgments and methodologies. Layton asserts that this autonomy is difficult to maintain in the face of corporate business practices. The (directly employed) engineer is a hopeless captive in the "belly of the corporate whale."
>
> *The Revolt of the Engineers* relates the failures of engineering groups to improve their situation through licensing and unionizing over the past two hundred years. This book is hard to find, but it's well worth reading if you can get your hands on it.

Jeffrey J. Mayer, *If You Haven't Got the Time to Do It Right, When Will You Find the Time to Do It Over?* New York: Simon and Schuster, 1990.

> Consultants and other busy professionals often find themselves overwhelmed with high-priority tasks. Yet they, like everybody else, have only twenty-four hours in the day and limited amounts of energy. Using your time wisely not only increases your income but also nurtures emotional competence and prevents burnout.
>
> Mayer's book is a short and breezy guide to time management. It shows you how to organize your work to be more efficient, how to plan and schedule your day, how to avoid time wasters, and how to overcome the "sidetrack syndrome." It teaches you how to organize your desk and files so that they work for you, not against you. Mayer helps you sort out the priorities and truly *manage* your time.

Michael J. McCarthy, *Mastering the Information Age: A Course in Working Smarter, Thinking Better, and Learning Faster.* Los Angeles: Jeremy P. Tarcher, 1991.

> McCarthy's well-written book on information mastery offers a variety of techniques to handle information overload. His approach differs from the "info surfing" I described in Chapter 16. McCarthy says the way to overcome information overload is to learn how to become a more powerful swimmer as you swim through the sea of information. Toward this end, he gives many interesting methods for speed-reading, improving critical thinking skills, enhancing memory, and proper care of the "machine" (your brain). By following this methodology, you can improve the "throughput" ability of your mind and master information overload.

C. W. Metcalf and Roma Felible, *Lighten Up: Survival Skills for People Under Pressure*. Reading, MA: Addison-Wesley, 1992.

> *Laughter is a reflex but unique in that it has no apparent biological purpose. One might call it a luxury reflex. Its only function seems to be to provide relief from tension.*
>
> —Arthur Koestler

Professional demeanor requires a committed and concerned attitude about being helpful to your clients and achieving your own goals. However, operating a consulting business can be very stressful if you approach it with an attitude that is too serious. *Lighten Up* offers a fun way to relax and avoid burnout. It shows how a relaxed attitude promotes creativity, resilience, and higher productivity. Laughter and lightness do have a place on the job. The authors describe humor as the antidote for "terminal professionalism." For them, humor is not telling jokes, but looking at things in perspective, valuing the joys of living, and learning not to take yourself—or others —too seriously.

John T. Molloy, *John T. Molloy's New Dress for Success*. New York: Warner Books, 1988.

Molloy gives valuable and straightforward advice on dressing for business. His comments on fit, quality, and fabric patterns are especially useful for the beginning consultant. Your visual image will not help with the technical aspects of your work, but it will make dealing with others easier. A few hours reading this book is time well spent.

Miyamoto Musashi, *A Book of Five Rings*. Woodstock, NY: The Overlook Press, 1982.

This Japanese classic on strategy is three hundred years old, but its material is timeless. The businessman as well as the warrior needs to understand when to attack, when to retreat, when to bluff, and when to wait. Most of us employ only a very limited number of strategies in our daily lives, regardless of their appropriateness to the moment. Musashi's theme is to expand your "bag" of strategies to accommodate more of the various situations you will encounter.

Gerald Nadler and Shozo Hibino, *Creative Solution Finding*. Rocklin, CA: Prima Publishing, 1995.

This book shows that traditional problem-solving methods are inadequate for many kinds of problems. The authors' new paradigm for problem solving, called *breakthrough thinking*, avoids many of the traps that result from rigidly following the traditional approach. Breakthrough thinking can be attained by following seven simple principles, including the Uniqueness Principle ("Assume initially that your problem is different than any other. Don't start by copying pre-existing solutions.") and the Systems Principle ("Everything is a system. Use a systems framework to identify the elements, dimensions, and interrelationships of your solution. Don't assume the details will work out.").

The authors have an earlier book called *Breakthrough Thinking* that covers much of the same material. However, *Creative Solution Finding* is more readable and slightly better organized.

Anthony O. Putman, *Marketing Your Services: A Step-by-Step Guide for Small Businesses and Professionals.* New York: John Wiley & Sons, 1990.
> Putman's superb guide to marketing is aimed specifically at small businesses, professionals, and consultants. He offers solid advice on how to define your services and align yourself with your market. This book emphasizes the idea that marketing consists of *building relationships with customers,* right from the very first contact. In Putnam's view, developing and maintaining professional relationships is the key to marketing success. Putman has a good sense of humor and gives many examples that bring his material to life.

Michael Ray and Rochelle Myers, *Creativity in Business.* Garden City, NY: Doubleday & Company, Inc., 1986.
> This book is based on the authors' creativity course at Stanford's Graduate School of Business. You can tell you're in for a treat when you see chapter titles like "If at First You Don't Succeed, Surrender," "Destroy Judgment, Create Curiosity," and "Ask Dumb Questions." *Creativity in Business* is a thought-provoking and entertaining introduction to becoming more creative in business. Recommended.

Anthony Robbins, *Awaken the Giant Within.* New York: Summit Books, 1991.
> *The definition of insanity is doing the same things over and over again and expecting a different result.*
>
> —Anthony Robbins

> We all know about being stuck in a rut (that is, not changing). Robbins's book is about changing: changing your thoughts, changing your emotional reactions, changing your habits, and changing your life into what you want it to be.
>   *Awaken the Giant Within* is based on Robbins's success seminars. It offers a potent mixture of psychology, positive thinking, and personal values exploration that will help you gain control over your future. This substantial book (540 pages) is not an anthology of "the little engine that could" platitudes, but a sound approach based on neurolinguistic programming and modern psychology. Robbins gives exercises to restore self-esteem, conquer self-criticism, and unleash the passion and drive to achieve your goals.

R. Robert Rosenberg et al., *Business Law: UCC Applications,* Sixth Ed. New York: McGraw-Hill, 1983.
> This classic business text discusses contracts, sales, commercial paper, insurance, and business organization. It has many clear, short examples and cross-references to the Uniform Commercial Code, which is the set of rules that govern commercial transactions in banking, contracts, and sales.

Jim Schneider, *The Feel of Success in Selling.* Englewood Cliffs, NJ: Prentice Hall, 1990.
> *Feeling successful is the first, and most important, step in selling. You can't lead a cavalry charge if you think you look funny sitting on a horse.*
>
> —Jim Schneider

This down-to-earth book is based on the sales training seminars given by the author. Although it is not directed at technical sales in particular, it offers many valuable techniques. The author focuses on the psychology of the sales transaction. He emphasizes centering on the customer's viewpoint and needs—not your own. This book is full of helpful advice for establishing rapport, uncovering needs, developing trust, overcoming resistance, and closing the sale. The author maintains that analyzing your mistakes isn't very useful in the long run. Instead, look at your successes, no matter how small. Once you know what you're doing right, "bottle" that feeling and use it to stimulate further success.

Charles "Chic" Thompson, *What a Great Idea: Key Steps Creative People Take.* New York: Harper Collins, 1992.

Creativity is the ability to look at the same thing as everyone else but to see something different.

—Chic Thompson

Thompson's book on creativity can help consultants in two ways. First, it can enhance your abilities as a professional problem solver, enabling you to offer fresh ideas, more options, and new solution methods to your clients. Second, it can encourage you to greater creativity in setting up and marketing your consulting practice. *What a Great Idea* offers methods to establish a creative environment, "map" your ideas so you can visualize their interconnections, nourish your vision, jump-start unproductive meetings, and fight "killer" phrases.

Denis Waitley, *The Psychology of Winning.* New York: Berkley Books, 1984.

Winning is taking the talent or potential you were born with, and have since developed, and using it fully toward a goal or purpose that makes you happy.

—Denis Waitley

This motivational classic teaches you how to be a winner. Through exhaustive research, Waitley was able to identify ten qualities of winners in all walks of life. Some of these are positive self-esteem, positive self-control, positive self-motivation, and positive self-image. Motivational books such as this are extremely important for consultants going into business for themselves. Breaking away from the pack means that you must develop your own business goals and learn how to generate the emotional energy to pursue them. *The Psychology of Winning* might help "prime the inspiration pump" in your annual planning sessions. (A good source for this and other motivational books, tapes, and videos is the Nightingale-Conant Corp. For a catalog, write to 7300 North Lehigh Ave., Niles, IL 60714, call at 800-525-9000, or visit the Nightingale-Conant Web site at http://www.nightingale.com.)